Alternative Fuels and Their Application to Combustion Engines

Alternative Fuels and Their Application to Combustion Engines

Editor

S M Ashrafur Rahman

MDPI • Basel • Beijing • Wuhan • Barcelona • Belgrade • Manchester • Tokyo • Cluj • Tianjin

Editor
S M Ashrafur Rahman
School of Mech., Medical
Process Engineering
Queensland University of
Technology
Brisbane
Australia

Editorial Office
MDPI
St. Alban-Anlage 66
4052 Basel, Switzerland

This is a reprint of articles from the Special Issue published online in the open access journal *Energies* (ISSN 1996-1073) (available at: www.mdpi.com/journal/energies/special_issues/alternative_fuel_engines).

For citation purposes, cite each article independently as indicated on the article page online and as indicated below:

LastName, A.A.; LastName, B.B.; LastName, C.C. Article Title. *Journal Name* **Year**, *Volume Number*, Page Range.

ISBN 978-3-0365-1398-0 (Hbk)
ISBN 978-3-0365-1397-3 (PDF)

Contents

About the Editor

S M Ashrafur Rahman

Dr S M Ashrafur Rahman is currently working as an Environmental Consultant at Trinity Consultants Australia. He is also working as a Visiting Fellow at Queensland University of Technology (QUT). He obtained his Doctoral degree in Mechanical Engineering from QUT in 2018. Previously, he completed his Master of Engineering Science from the University of Malaya in 2015 and Bachelor of Science in Mechanical Engineering from Bangladesh University of Engineering and Technology in 2012. His research work focuses on evaluating various bio-oils as possible supplements to petro-diesel. In the last 8 years, he has published 56 ISI indexed journals with a total citation of 2632. He is an esteemed reviewer of several WOS-indexed journals.

Article

Overview of Sustainable Aviation Fuels with Emission Characteristic and Particles Emission of the Turbine Engine Fueled ATJ Blends with Different Percentages of ATJ Fuel

Paula Kurzawska * and Remigiusz Jasiński

Faculty of Civil and Transport Engineering, Poznan University of Technology, 60-965 Poznan, Poland; remigiusz.jasinski@put.poznan.pl
* Correspondence: paula.kurzawska@put.poznan.pl

Abstract: The following article focuses on sustainable aviation fuels, which include first and second generation biofuels and other non-biomass fuels that meet most of environmental, operational and physicochemical requirements. Several of the requirements for sustainable aviation fuels are discussed in this article. The main focus was on researching the alcohol-to-jet (ATJ) alternative fuel. The tests covered the emission of harmful gaseous compounds with the Semtech DS analyzer, as well as the number and mass concentration of particles of three fuels: reference fuel Jet A-1, a mixture of Jet A-1 and 30% of ATJ fuel, and mixture of Jet A-1 and 50% of ATJ fuel. The number concentration of particles allowed us to calculate, inter alia, the corresponding particle number index and particle mass index. The analysis of the results made it possible to determine the effect of the content of alternative fuel in a mixture with conventional fuel on the emission of harmful exhaust compounds and the concentration of particles. One of the main conclusion is that by using a 50% blend of ATJ and Jet A-1, the total number and mass of particulate matter at high engine loads can be reduced by almost 18% and 53%, respectively, relative to pure Jet A-1 fuel.

Keywords: alcohol-to-jet; alternative fuel; SAF; emission; particles; particulate matter

 **check for updates**

Citation: Kurzawska, P.; Jasiński, R. Overview of Sustainable Aviation Fuels with Emission Characteristic and Particles Emission of the Turbine Engine Fueled ATJ Blends with Different Percentages of ATJ Fuel. *Energies* **2021**, *14*, 1858. https://doi.org/10.3390/en14071858

Academic Editor: S M Ashrafur Rahman

Received: 3 February 2021
Accepted: 23 March 2021
Published: 26 March 2021

Publisher's Note: MDPI stays neutral with regard to jurisdictional claims in published maps and institutional affiliations.

1. Introduction

Along with the development of the aviation industry, its share in the emission of environmental pollution has increased. Currently, the aviation industry is responsible for an estimated 2% of global greenhouse gas emissions [1,2]. In 2015, aviation operations generated over 781 million tons of carbon dioxide, and it is expected that by 2050, based on forecasts of air traffic growth, 2700 million tons of carbon dioxide will be generated annually [2]. According to forecasts by Airbus, air traffic doubles every 15 years and the number of flights increased by 80% between 1990 and 2014 [3,4]. The growing number of air connections unfortunately significantly affects environmental pollution and the associated climate change effects. As a result, many aviation organizations and airlines are taking measures to reduce greenhouse gas emissions from the aviation industry, including using alternative fuels. The increase in interest in alternative fuels is caused not only by climate change and the impact of burning conventional fuels on the environment, but also by the depletion of natural resources of crude oil [5], rising oil prices and countries' dependence on their suppliers. The aviation sector wants to ensure security of supply alternative aviation fuels at affordable prices [6]. Alternative fuels obtained from plants or other raw materials would achieve energy independence from Organization of Petroleum Exporting Countries (OPEC) member states, whose political instability is associated with frequent changes in oil prices [7]. For this reason, more and more countries and airlines are investing in research to produce sustainable aviation fuels from alternative sources, e.g., used oil, municipal waste, algae or even plastic. In 2018 the Renewable Energy Directive II (REDII) entered into force, which increased targets for the share of renewable fuels in transport from 10% by

2020, to 14% by 2030 [8]. Also due to REDII savings of greenhouse gas emissions from use of renewable liquid and gaseous fuels made from non-biological origin in transport field shall be at least 70% from the year 2021 [9]. In the aviation field there are currently seven approved technologies for the production of alternative aviation fuels, which include for example hydroprocessed esters, hydroprocessed fermented sugar and alcohols. Alcohols have a huge potential as alternative fuels, because of their liquid nature, production from renewable biomass and high oxygen contents and also high cetane number. Fuels which contain oxygen can reduce the combustion chamber parameters, like temperature, and through this emission of harmful gaseous compounds can be reduced [10]. Therefore, the work below focuses on the study of the concentration of harmful exhaust compounds and particles in the engine exhaust, depending on the degree of mixing of the alternative alcohol-to-jet (ATJ) fuel with conventional Jet A-1 fuel.

2. Sustainable Aviation Fuels

Sustainable aviation fuels (SAF) is the principal term used to refer to non-conventional aviation fuels. Another names are sustainable alternative fuel, biojet or renewable jet fuel [11]. The term sustainable aviation fuels covers not only biofuels, but also fuels produced from raw materials other than biomass, such as waste. Biofuels refer to fuels produced from raw materials of plant or animal origin, and due to their aggregate state, we divide them into solid, liquid and gas [11]. In the aviation industry, biofuels mainly refer to liquid biofuels [2]. In order to qualify as "sustainable" aviation fuels must meet the following criteria [11]:

- Reducing carbon dioxide emissions throughout the life cycle;
- Limited need for fresh water;
- No need for deforestation and no competition with food production for land for cultivation.

Biofuels used in aviation can be divided into first, second and third generation biofuels according to the general division of biofuels. In this analysis of alternative fuels, the 1st generation fuels have been omitted due to the fact that they cannot be called sustainable fuels, as their production uses food crops [12]. Second-generation biofuels are fuels obtained from inedible plants or plant waste, which can be grown on less fertile soils, and even on wastelands [13]. This group includes wood and its waste, which contain lignocellulosic biomass, organic waste and food waste from agri-food processing [14]. They do not compete with food cultivation as they come from a separate biomass, but some biomass still competes with land use, as it grows in the same climate as food crops [15]. Other raw materials, which are not biomass, are currently in the phase of physicochemical research and testing, or test flights are being carried out with their use. Most often it is waste from households and companies. Research on the use of clothes, bottles, leftovers and newspapers has also been started. The use of municipal solid waste (MSW) has a very large potential, due to the use of raw materials that would be stored and would emit carbon dioxide, and thanks to re-use they can drive aircraft engines [13]. These are e.g., fuels produced from municipal waste.

The second-generation raw material is jatropha oil. It is sourced from jatropha seeds, which are poisonous to both humans and animals. 30 to 40% of the seed weight can be obtained from each seed. Jatropha has low soil and climatic requirements, therefore it can be cultivated in difficult conditions, such as dry and undeveloped areas [14]. As a result, it does not compete with food crops for arable land. Jatropha is subjected to the process of oil extraction, which produces bio oil, and then it is treated with hydrogen to obtain a fuel of the hydroprocessed renewable jet (HRJ) type [16]. Another oilseed plant is camellina. It is often cultivated as a crop rotation plant, so like jatropha—it does not compete with food crops for arable land [13]. It occurs mainly in a temperate climate, in Central Europe, Finland and the United States [17]. The latter is also subjected to an oil extraction process and then treated with hydrogen to obtain HRJ fuel.

Vegetable and animal oils, which are already waste and will not be used further in the food industry, can also be considered as second-generation biofuels. Used vegetable oils can be treated with hydrogen to make jet fuel. It is currently one of the most promising raw materials for the production of alternative aviation fuels.

Aviation biofuels are processed differently, depending on the raw material used. The specific group of raw materials and the corresponding transformations are shown in Figure 1. This article focuses on alcohol-to-jet fuel, made from starch and sugar crops.

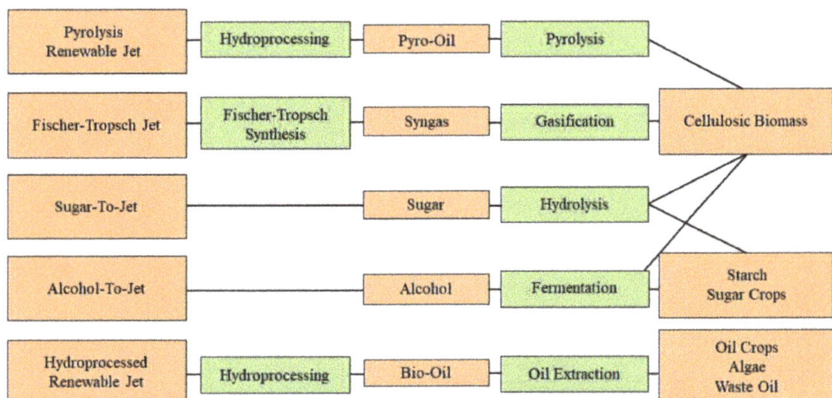

Figure 1. Raw materials used for the production of aviation biofuels and the corresponding processing [17].

A renewable fuel option for aviation is also power-to-liquids (PtL) production pathway, which is based on electricty, water and carbon dioxide. The first step in PtL is electrolysis of water in which hydrogen is producted from renewable electricity. Afterwards carbon dioxide is supplied and and at the last step is synthesis to liquid hydrocarbons with subsequent conversion to refined fuels. Power to liquid can use renewable electricity and CO_2, for example from biomass or from the air. What's the most important, PtL fuels can be close to carbon neutral and need less water than several biogenic fuels, for example from jathropha plants, which need a lot of water to grow. Water in Pt: technology is needed as a hydrogen source, but they amount needed is still less than for growing some of the plants used in production of biofuels [18,19].

3. Requirements for Alternative Aviation Fuels

3.1. Operational Requirements

One of the main requirements for alternative aviation fuels is their compatibility with the existing fuel infrastructure. This includes the pipelines through which fuel is transported, refueling systems for the aircraft and the engine structure itself [20]. It is very important that the change of conventional fuel to alternative fuel does not require changes in design and infrastructure, as this would significantly hinder the entry of alternative fuels into the aviation market. Therefore, an ideal sustainable aviation fuel would be 100% compatible in operation with currently used aviation fuels [20,21]. Such fuel is known as "drop-in" [13]. There is an alternative fuel compatibility assessment with existing infrastructure, which assigns a neutral assessment if the fuel is fully compatible and does not require any interference with the existing system, and a negative assessment if the fuel requires a complete system change [20]. For now, only alternative and conventional fuels may be mixed.

3.2. Physicochemical Requirements

Aviation fuels must meet a number of requirements regarding their physicochemical properties in order to be used in aircraft engines. Physicochemical properties of aviation fuels are the main determinant of safe flight performance, therefore they must be strictly observed. The ASTM D1655 standard specifies the specific values of the physicochemical properties of aviation fuels. For alternative aviation fuels was assigned the standard ASTM D7566 [21], shown in Table 1.

Table 1. Physicochemical properties according to ASTM D7566 standard.

No.	Property Name	Unit of Measure	Requirements acc. to ASTM D7566
1	Density at 15 °C	kg/m^3	from 775 to 840
2	Viscosity at −20 °C	mm^2/s	max 8.0
3	Viscosity at −40 °C	mm^2/s	max 12
4	Calorific value	MJ/kg	min 42.8
5	Aroma content	%	min 8, max 25
6	Naphthalene content	%	max 3.0
7	Flash-point	°C	min 38
8	Crystallization temperature	°C	max −47
	Distillation:		
	Start distillation temperature	°C	-
9	10% distils to temperature	°C	max 205
	End distillation temperature	°C	max 300
	Residue	%	max 1.5
	Loss	%	max 1.5
10	Lubricity	mm	max 0.85

This standard specifies the maximum share of alternative fuels in the mixture consisting of conventional and sustainable fuels at the level of 50% in volume terms. At least half of the mixture must be Jet A or Jet A-1 fuel [13]. Alternative fuels that meet the requirements of ASTM D7566 can be used in aircraft engines that require the D1655 aviation fuel standard [22]. The approved aviation fuel production methods are presented in Table 2.

Table 2. Methods of producing alternative fuels approved by the ASTM D7566 standard (own study based on [4,14,23,24]).

Annex	Process	Raw Material	Approval Date	Blending Limit
A1	Fischer-Tropsch Synthetic Paraffinic Kerosene (FT-SPK)	Biomass (wood waste, grass, municipal solid waste)	2009	up to 50%
A2	Hydroprocessed Esters and Fatty Acids (HEFA-SPK)	Oily biomass, e.g., algae, jatropha, camelina	2011	up to 50%
A3	Hydroprocessed Fermented Sugars to Synthetic Isoparaffins (HFS-SIP)	Bacterial conversion of sugars into hydrocarbons	2014	10%
A4	FT-SPK with aromatics (FT-SPK/A)	Renewable biomass, i.e., municipal solid waste, agricultural and wood waste	2015	up to 50%
A5	Alcohol-to-jet Synthetic Paraffinic Kerosene (ATJ-SPK)	Agricultural waste (corn shoots, grass, straw), cellulosic biomass	2016	up to 50%
A6	Catalytic Hydrothermolysis Synthesized Kerosene (CH-SK, or CHJ)	Vegetable or animal fats, oils and greases	2020	up to 50%
A7	Hydroprocessed Hydrocarbons, Esters and Fatty Acids Synthetic Paraffinic Kerosene (HHC-SPK or HC-HEFA-SPK)	Hydrocarbons of biological origin, fatty acid esters, free fatty acids, or a species of *Botryococcus braunia* algae	2020	up to 10%

3.3. Environmental Requirements

One of the main reasons for studying alternative fuels and looking for new solutions to power jet engines is the impact of crude oil and its derivatives on the environment and climate change. Emissions of harmful compounds are related to the physical and chemical properties of fuel [24]. High emissions of carbon dioxide, greenhouse gases and other harmful substances generated during the combustion of conventional aviation fuels, such as carbon oxides, nitrogen oxides, hydrocarbons and particles, increase the interest in alternative fuels. Carbon dioxide absorbed by plants during the growth of biomass is similar to the amount of carbon dioxide emitted during the combustion of fuel from that biomass, which makes it possible to remain neutral in terms of greenhouse gas emissions [13]. Sustainable aviation fuels should enable a significant reduction in greenhouse gas emissions during their combustion, but also, which is crucial, throughout their entire life cycle, from the growth and fertilization of plants and algae, through their transport, processing, distribution, and end use in the engine aviation. Life cycle emissions are mainly related to second and third generation biofuels which are based on plants and algae that need to be grown for use in the aerospace industry.

By using fuel based on wood biomass, 95% of CO_2 can be saved compared to the currently used jet fuel. Wood biomass is one of the lowest carbon dioxide emissions per MJ of fuel, only algae fuel has a greater one, which in a realistic case could be 98% greenhouse gas emissions, and in the best scenario up to 124%, compared to conventional jet fuel. This is due to the fact that during their growth and development, algae absorb large amounts of carbon dioxide, which in the case of their total CO_2 emission, may be below zero. Carbon dioxide emissions for other alternative fuels obtained e.g., from conventional oil, jatropha or animal fuels range from 20% to 90% depending on the raw material used, with the least preferred fuel being from oil plants. The above data is indicative of the fact the specific emission value for each of the analyzed fuels depends on the method used for producing the alternative fuel [25].

4. Experimental

The fuels used during the research were the alternative fuel alcohol to jet synthesized paraffinic kerosene (ATJ-SPK) from isobutanol and the comparative fuel Jet A-1. The alternative fuels supplying the engine during the tests were mixed in the following volume proportions with conventional JetA-1 fuel: 30% ATJ fuel and 50% ATJ fuel. During the tests, the concentration of carbon oxides (CO), carbon dioxide (CO_2), hydrocarbons (HC) and nitrogen oxides (NO_x) was measured, as well as the concentration of the number of particles by particle diameter.

The conventional fuel Jet A-1 is produced during the fractional distillation of crude oil, known as rectification. It is aviation kerosene, i.e., the liquid fraction of distilling crude oil ranging from 130 °C to about 280 °C. Due to the low octane number and simple production technology, it is relatively cheap—cheaper than gasoline or diesel. Jet A-1 is used in civil aviation, and it differs from Jet A mainly in the freezing point, which is −47 °C for Jet A-1 and −40 °C for Jet A [26].

The ATJ alternative fuel can be produced by many different conversion routes, but each starts with a biomass feedstock. The raw materials used to produce ATJ fuel are, for example, sugar cane, sugar beet, cereals or lignocellulosic biomass. ATJ fuel is made by converting alcohols such as methanol, ethanol, butanol and long-chain fatty alcohols. The maximum use of ethanol in the production of ATJ fuel is 10–15%. ATJ fuels currently used in aviation and meeting the requirements of ASTM D7566 are isobutanol- and ethanol-based fuels, however Annex A5 (ATJ-SPK) is ultimately to include the use of any alcohols containing from two to five carbon atoms [25,27–29]. Use of ATJ fuel in an aircraft engine requires a maximum of 50% ATJ blend with conventional fuel. The process of converting alcohol into alcohol-to-jet fuel includes the following processes: isobutanol or ethanol dehydration, oligomerization, hydrogenation and fractionation to obtain a component which is a mixture of hydrocarbon jet fuel (Figure 2) [23].

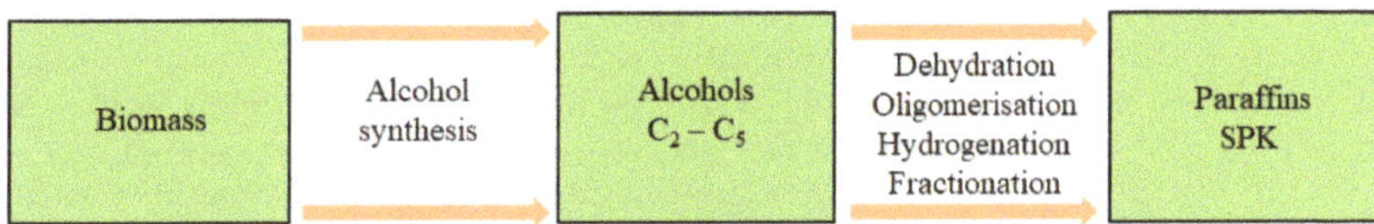

Figure 2. Simplified diagram of the ATJ-SPK process [30].

According to the European Aviation Environmental Report 2019 prepared by the European Union Aviation Safety Agency (EASA), thanks to the use of ATJ fuel, the percentage reduction in greenhouse gas emissions compared to the use of conventional fuel ranges from 26 to 74%, depending on the raw material used in the production. These estimates do not take into account the greenhouse gas emissions in the raw material growth phase. In the case of ATJ aviation fuel based on isobutanol, the lowest percentage reduction can be achieved by using maize grain as a raw material (54%), and the highest by using forest residues (74%). On the other hand, in the case of ATJ fuel obtained from ethanol, the lowest reduction in greenhouse gas emissions can be obtained by using corn kernel as a raw material (26%), and the highest—sugar cane (69%). Table 3 presents the physicochemical properties of the tested fuels [4].

Table 3. Physicochemical properties of tested fuels [own study based on [31,32].

Property	Jet A-1 [According to ASTM D1655 Standard]	ATJ-SPK	50/50% v ATJ-SPK and Jet A-1
Crystallization temperature [°C]	−47	−61	−54
Flash point [°C]	min 38	48	min 38
Calorific value [MJ/kg]	42.8	43.2	43.8
Total sulfur content [%]	max 0.3	<0.01	0.02
Aromas content [%]	17.3	0	8.8

The tests were carried out on a GTM-120 miniature turbine engine, made of a centrifugal compressor, diffuser, annular combustion chamber with pre-vaporising tubes, turbine nozzle, turbine wheel and nozzle cone. The engine is started by a starter. The technical parameters of the described engine are presented in Table 4.

Table 4. Technical parameters of the GTM-120 engine.

Maximum thrust [N]	100
Fuel consumption (for maximum thrust) [g/min]	520
Length [mm]	340
Width [mm]	115
Weight [kg]	1.5

The Semtech DS analyzer from Sensors Company (city, state abbrev if USA, country) was used to measure the concentration of gaseous exhaust compounds. This analyzer measures the concentration of nitrogen oxides, hydrocarbons, carbon monoxide and carbon dioxide. The exhaust gases from the GTM-120 engine were fed to the analyzer via a probe placed 3 cm from the outlet nozzle and a cable with a temperature of 191 °C, required to measure the concentration of hydrocarbons in the flame ionization analyzer. Then after cooling the exhaust gases to the temperature of 4 °C, measurements of the concentrations of carbon monoxide, carbon dioxide and nitrogen oxides were carried out. The Semtech DS analyzer includes the following measurement modules [32]:

- A flame ionization detector (FID), which uses the change of electric potential resulting from the ionization of molecules in the flame; it is used to determine the total concentration of hydrocarbons,
- A non-dispersive ultraviolet (NDUV) analyzer that uses ultraviolet radiation to measure the concentration of nitrogen oxide and dioxide
- A non-dispersive infrared (NDIR) analyzer using radiation infrared to measure the concentration of carbon monoxide and dioxide, and
- An electrochemical analyzer for determining the oxygen concentration in the exhaust gas.

At the same time, the particle number concentration was measured using an EEPS 3090 (Engine Exhaust Particulate Sizer™ spectrometer) analyzer from TSI Incorporated (city, state abbrev if USA, country). This analyzer measures the discrete range of particle diameters from 5.6 nm to 560 nm [32,33]. The exhaust gas was directed to the analyzer through a dilution system, where the total flow was 10 l/min, including the exhaust gas flow 0.3 l/min, so the exhaust gas in the tested sample accounted for 3%. Technical data of the EEPS 3090 analyzer are presented in Table 5.

Table 5. Technical data of the EEPS 3090 analyzer [33].

Parameters	Value
Diameter of the measured particles	5.6–560 nm
Number of measurement channels	16 channels per decade
Resolution	10 Hz
Exhaust sample volume flow rate	0.6 m^3/h
Compressed air volume flow rate	2.4 m^3/h
Input sample temperature	10–52 °C

The measuring range was from 10 to 100 N, and the measurements were made every 10 N. In order to clearly present the results, after the measurements, the measuring range was reduced to the following three ranges: low engine operation load from 10 to 30 N, medium engine operation load from 40 to 60 N and high engine operation load ranging from 70 to 100 N (Table 6). Measurement results were averaged in each of the examined areas.

Table 6. Measurement ranges for engine operation load and their values.

Range Name	Engine Operation Load Range [N]	
	Up	To
Low	10	30
Medium	40	60
High	70	100

5. Results and Discussion

5.1. Concentration of Harmful Exhaust Gas Compounds

The results were grouped for easier comparison by reference to the fuel composition and the engine load range. During the measurements, the focus was on the concentration of harmful exhaust gas compounds, such as CO_2, CO, HC and NO_x. The measurement results of the tested exhaust gas compounds are shown in Figure 3. It is worth underlining that the maximum thrust power of the GTM 120 engine, amounting to 100 N, was achieved only with the use of pure conventional Jet A-1 fuel. For a fuel containing 30% and 50% ATJ fuel the maximum load was about 90 N. Carbon dioxide, carbon monoxide, hydrocarbons and nitrogen oxides are the main products of combustion. Emissions of these harmful gaseous exhaust compounds depend on the engine load, so also on flight mode. Basically, emission of carbon dioxide is proportional to fuel consumption. Emission of carbon monoxide is high for low engine load, for example for idling and taxiing, and decreases when engine load is increasing. The opposite situation is true for nitrogen oxides. NOx emissions increase with increasing engine load, so they are high for climbing and take-off. Nitrogen oxides and unburned hydrocarbons are formed, inter alia, depending on the temperature and pressure in the engine [34,35].

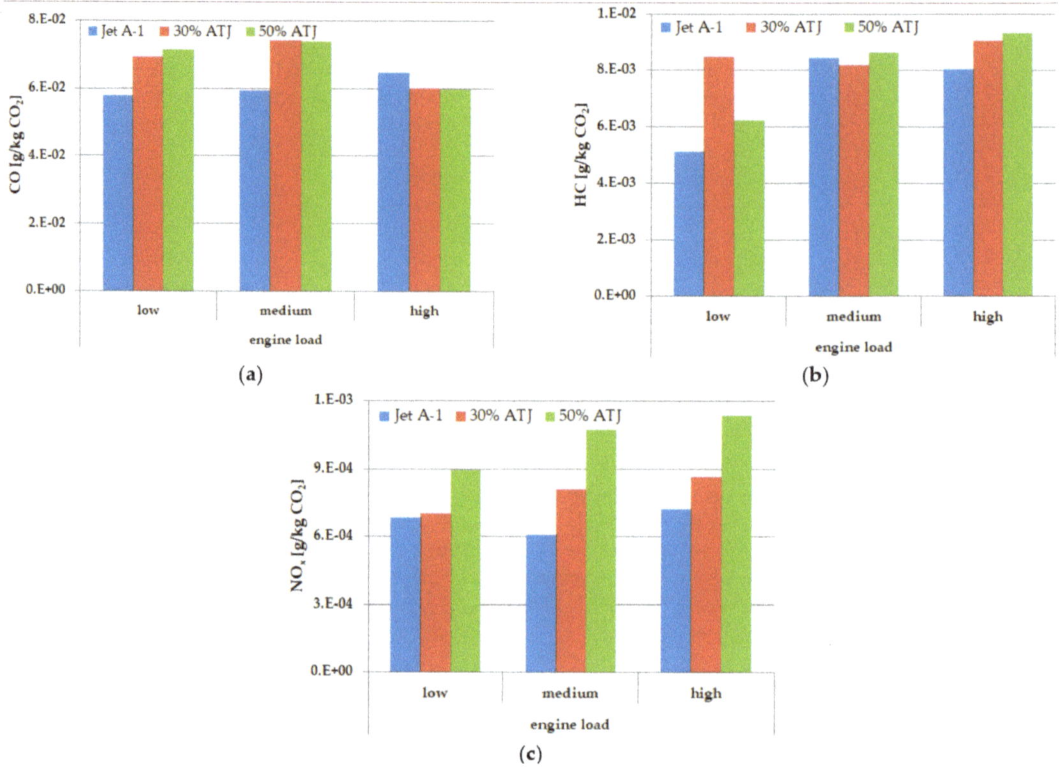

Figure 3. Emission index for (**a**) carbon monoxide; (**b**) hydrocarbons; (**c**) nitrogen oxides.

The emission of harmful gaseous compounds was presented in the form of emission factors related to the emission of carbon dioxide during each measurement. Comparing the emissions of carbon monoxide (Figure 3a) with CO_2 emissions at low and medium engine operation load, it was found that it is the lowest for pure Jet A-1 fuel, while in the case of high engine operation load for Jet A-1 fuel it is the highest, compared to other tested fuels. Carbon monoxide emissions for between 30% and 50% ATJ fuel do not differ significantly over the entire operating range of the engine. The difference between Jet A-1 and 50% ATJ for low engine operation load is 23% and for medium engine operation load it is 25%, when the emission was higher for 50% ATJ. In turn comparing high engine operation load, emissions of CO related to CO_2 were the lowest for 50% ATJ and about 8% lower than for Jet A-1.

On the other hand, in the case of hydrocarbon emissions (Figure 3b), for low engine operation load, the highest HC emission is shown for fuel with 30% ATJ content, while at medium and high engine operation load it is for fuel with 50% ATJ content. For low engine operation load the difference between 30% ATJ and Jet A-1 is 40% and between 30% ATJ and 50% ATJ is 36%. In turn for high engine operation load difference between the highest emission for 50% ATJ and Jet A-1 is 16% in favor of the Jet A-1 fuel.

Comparing the emission of nitrogen oxides (Figure 3c), it was found that the lowest emissions occur for the pure conventional fuel Jet A-1 in the entire engine operation range, while the highest carbon oxide emission per CO_2 emission occurs for the fuel with 50% ATJ content in the entire engine operation range. The differences between Jet A-1 and 50% ATJ are respectively 31% for low load engine operation, 76% for medium load and 57% for high load engine operation. In this case, increasing the content of alternative fuel Alcohol-to-Jet in the mixture of Jet A-1 and ATJ is expected to increase the emission of nitrogen oxides.

5.2. Particles Concentration

Based on the obtained data for the particle number concentration, the characteristics of the mass concentration of particles depending on their diameter were calculated. For the calculation the solids density characteristic was used (Figure 4), which decreases with increasing particle diameter. The particle density function was determined empirically on the basis of the CFM56-7B26/3 aviation engine [36–38]. Knowing the diameter of particles, it was possible to calculate the mass of particulate matter by using the density and volume of the particles.

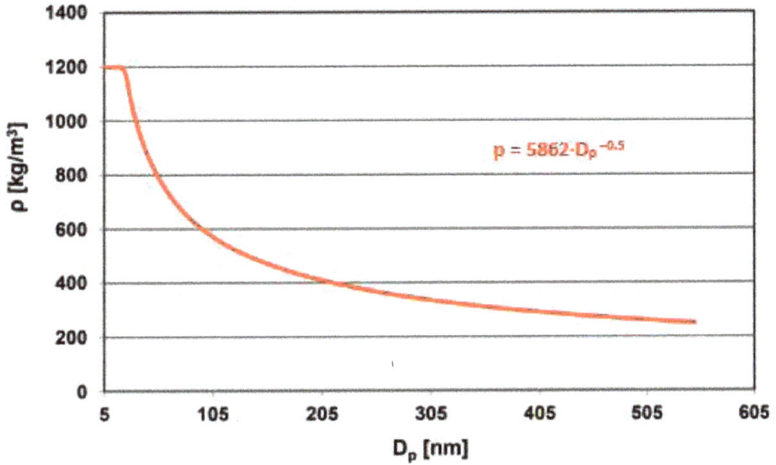

Figure 4. Density of solid particles depending on their diameter [32].

5.2.1. Low Engine Load

Figure 5 shows the number and mass concentration of particles depending on their diameter (first column) and cumulative values of the relative particle number and relative mass of particulate matter (second column) for low load engine operation fueled Jet A-1 (first row), 30% ATJ (second row) and 50% ATJ (third row). The cumulative curves were determined by standardizing the obtained data for the number and mass of particulate matter to the value 1 and using the connection between the quotient of the number of particles and the maximum number of particles as well as the quotient of the mass of particles and the maximum mass of particles.

In the case of particles in the exhaust of an engine running on clean fuel Jet A-1 (Figure 5a), for low engine operation load, particles with 25.5–124.1 nm diameter dominated. The characteristic diameter, i.e., the highest number of particles, of the discussed number concentration characteristic was about 60.4 nm. Based on the characteristics of the mass concentration of particulate matter for Jet A-1 fuel at low engine operation load, the vast majority of the mass of particulate matter was due to particles with a diameter of 25.5–220.7 nm. The remainder of the particulate mass results from the emission of a very small amount of particulate matter with diameters in the 294.3–523.3 nm range. At low engine operation load fueled Jet A-1 (Figure 5b) the cumulative values of the relative particles number and relative mass of particulate matter show that 90% of the relative number of all particles emitted corresponds to 60% of their relative mass. About 90% of all particles are less than 100 nm in diameter.

In the case of low engine operation loads ranging from 10 N to 30 N for the fuel with 30% ATJ content (Figure 5c), the particle diameters of 25.5–107.5 nm dominated, and the majority were the particles with a diameter of 52.3 nm. Compared to the total particles concentration for Jet A-1, the 30% ATJ fuel had a slight reduction in total particle count. The main part of the emitted mass was particulate matter with diameters in the range 34.0–165.5 nm. The remainder of the particulate mass results from the emission of a very small number of particulate matter with diameters in the range 294.3–523.3 nm. Cumulative values of the relative number and relative mass of particles for a 30% ATJ fuel (Figure 5d) shows, that 90% of the relative number of all particles emitted is about 55% of their relative mass. About 90% of all particulate matter is less than 80 nm in diameter.

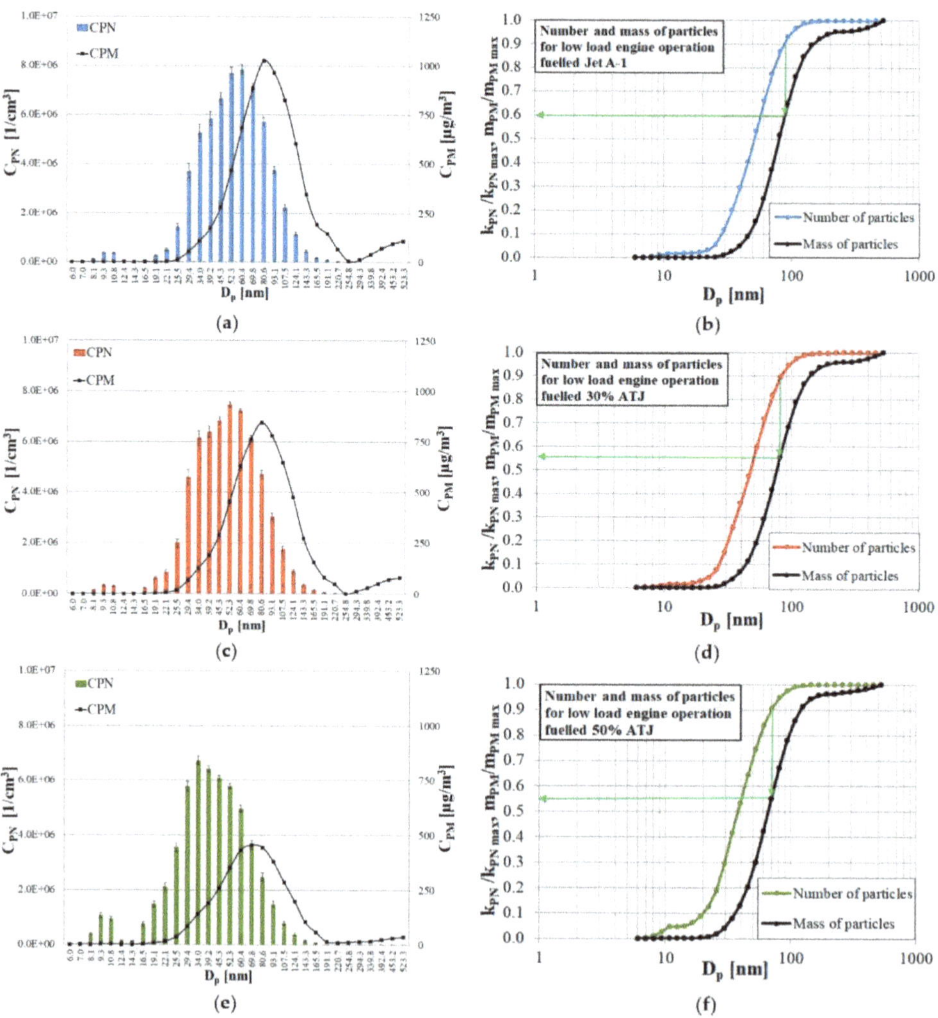

Figure 5. Number and mass concentration of particles depending on their diameter (**a,c,e**) and cumulative values of the relative particles number and relative mass of particulate matter (**b,d,f**) for low engine operation load fueled Jet A-1 (**a,b**), 30% ATJ (**c,d**) and 50% ATJ (**e,f**).

On the other hand, for the fuel containing 50% ATJ in the range of low engine operation load, the range of dominant particle diameters for low engine loads was 19.1–93.1 nm, and most of the particles had a diameter of 34.0 nm (Figure 5e). However, the concentration of the total number of particles in this case was the lowest compared to other tested fuels. The difference is approximately 10% compared to the total particles concentration for conventional Jet A-1. In the case of mass concentration of particulate matter, depending on their diameter, the main part of the emitted mass of particulate matter were particles with a diameter of 25.5–165.5 nm. The remainder of the particulate mass results from the emission of a very small number of particulate matter with diameters ranging from 254.8–523.3 nm. The total particulate mass concentration for the fuel containing 50% ATJ was 51% lower than the total particulate mass concentration for the Jet A-1 fuel in the discussed engine operating range. For a fuel containing 50% ATJ (Figure 5f), cumulative values of the relative number and relative mass of particles shows that 90% of the relative number of all particulate matter emitted is about 55% of their relative mass. About 90% of all particulate matter is less than 70 nm in diameter.

5.2.2. Medium Engine Load

Figure 6 shows number and mass concentration of particles depending on their diameter (first column) and cumulative values of the relative particles number and relative mass of particulate matter (second column) for medium engine operation load fueled Jet A-1 (first row), 30% ATJ (second row) and 50% ATJ (third row). In the case of medium load engine operation in the range of 40–70 N, for the conventional fuel Jet A-1, particles with a diameter of 25.5–107.5 nm were dominant, and the characteristic diameter was 52.3 nm (Figure 6a). Particles with a diameter of 29.4–165.5 nm accounted for the main share in the emitted mass of particles. The remainder of the particulate mass is due to the emission of a very small number of particles with diameters in the range 339.8–523.3 nm. Cumulative values of the relative number and relative mass of particles (Figure 6b) show that 90% of the relative number of all emitted particles is 60% of their relative mass. About 90% of all particulate matter is less than 80 nm in diameter.

For the fuel with 30% ATJ content, the particles with a diameter of 22.1–93.1 nm dominated for medium engine operation load, and the particles with a diameter of 34.0 nm were the most numerous (Figure 6c). As with the low engine operation load, the total particle concentration was lower than that of Jet A-1. In the case of mass concentration of particles, depending on their diameter, particles with a diameter of 25.5–124.1 nm accounted for the main share in the emitted particulate matter mass.

The remainder of the particulate mass results from the emission of a very small number of particles with diameters in the range 294.3–523.3 nm. Cumulative values of the relative number and relative mass of particles (Figure 6d) show that 90% of the relative number of all emitted particles is 60% of their relative mass. About 90% of all particulate matter is less than 70 nm in diameter.

In the case of the fuel with 50% ATJ content, for medium engine operation load, particles with a diameter of 16.5–80.6 nm dominated, however, particles with a diameter of 9.31 and 10.8 nm also appeared in a greater number (Figure 6e). The characteristic diameter was 34.0 nm. As with the low engine operation load, the total particles concentration was lowest for the fuel containing 50% ATJ compared to the fuel containing 30% ATJ and Jet A-1 fuel. The difference between Jet A-1 and 50% ATJ is less than 14%. Particles with a diameter of 25.5–124.1 nm were the main share in the emitted mass of particulate matter. The total particulate mass concentration was also the lowest compared to previous fuels in this engine load range and was about 40% of the total mass concentration of Jet A-1 for medium operation load. Cumulative values of the relative number and relative mass of particles (Figure 6f) show that 90% of the relative number of all emitted particles is also 60% of their relative mass. About 90% of all particles is less than 60 nm in diameter, so for medium load engine operation, when the fuel contained more ATJ alternative fuel, the diameter of 90% of all particles was smaller: for pure Jet A-1 it was 80 nm, for 30% ATJ—70 nm, and for the highest content of alternative fuel which was 50%, it was 60 nm.

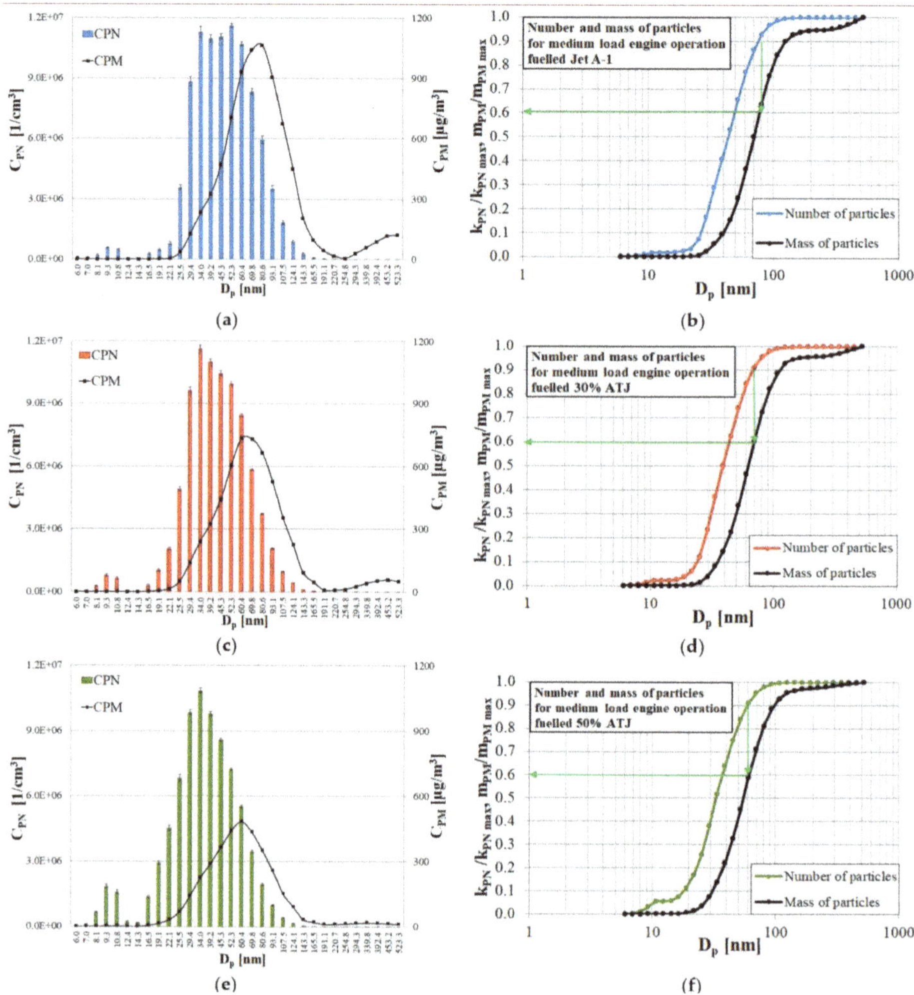

Figure 6. Number and mass concentration of particles depending on their diameter (**a,c,e**) and cumulative values of the relative particles number and relative mass of particulate matter (**b,d,f**) for medium engine operation load fueled Jet A-1 (**a,b**), 30% ATJ (**c,d**) and 50% ATJ (**e,f**).

5.2.3. High Engine Load

At high engine operation load (Figure 7), for each of the tested fuel, an increase in the concentration of the total number of particles was found. In the case of Jet A-1 fuel, particles with a diameter of 16.5–107.5 nm dominated for high engine load (Figure 7a). As can be seen, with increasing engine load, the diameter of the dominant particles decreased. For low and medium engine operation load, the most were particles with a diameter of 60.4 nm, while for high engine operation load, the most were particles with a diameter of 34.0 nm. However, the total number of particles is incomparably the highest at high engine operation load. On the other hand, the main part of the emitted mass of particulate matter were particles with diameters of 25.5–143.3 nm, and the remaining part of the mass of particulate matter results from the emission of a very small number of particles with diameters in the range of 294.3–523.3 nm. Cumulative values of the relative number and relative mass of particles (Figure 7b) show that 90% of the relative number of all emitted particles is also 60% of their relative mass. About 90% of all particles is less than 70 nm in diameter.

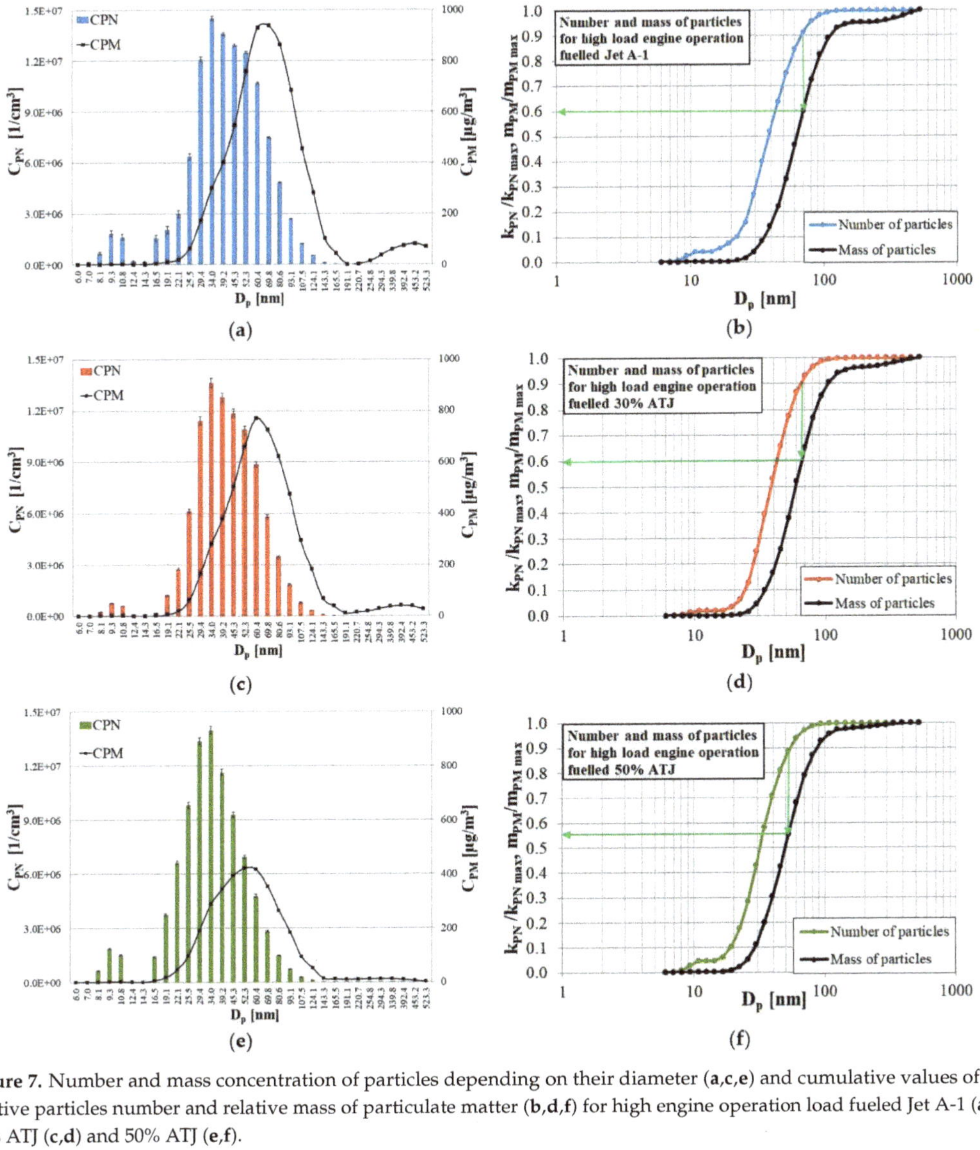

Figure 7. Number and mass concentration of particles depending on their diameter (**a**,**c**,**e**) and cumulative values of the relative particles number and relative mass of particulate matter (**b**,**d**,**f**) for high engine operation load fueled Jet A-1 (**a**,**b**), 30% ATJ (**c**,**d**) and 50% ATJ (**e**,**f**).

For the fuel with 30% ATJ fuel content, particles with a diameter of 22.1–93.1 nm dominated at high engine operation load (Figure 7c), and the most were particles with a diameter of 34.0 nm, similar to the medium engine operation load. In the case of the particulate matter mass concentration, depending on their diameter, the main fraction of the emitted particulate matter mass were particles with a diameter of 25.5–143.3 nm. The remainder of the particulate matter mass results from the emission of a very small number of particles with diameters in the 254.8–523.3 nm range. Cumulative values of the relative number and relative mass of particles (Figure 7d) show that 90% of the relative number of all emitted particles is also 60% of their relative mass. About 90% of all particles is also less than 70 nm in diameter.

On the other hand, when analyzing the numerical and mass concentration of particles for a 50% mixture of ATJ and Jet A-1 fuel (Figure 7e), it was found that in this case the numerical distribution of particles depending on the particle diameter in the entire engine operating range differs the most compared to the previous fuels. The range of diameters of the dominant particles for high loads was 16.5–80.6 nm, but there were also more particles with a diameter of 9.31 and 10.8 nm. The characteristic particle diameter was 34.0 nm. The total particle number concentration was approximately 18% lower than the total particle number concentration for conventional Jet A-1 fuel in the described engine operating range. In turn, the main part of the emitted mass of particulate matter were particles with a diameter of 22.1–124.1 nm. The total particulate mass concentration was 51% of the total Jet A-1 fuel mass concentration at the high engine operation load. Cumulative values of the relative number and relative mass of particles (Figure 7f) show that 90% of the relative number of all emitted particles is also 60% of their relative mass. About 90% of all particles is less than 55 nm in diameter.

5.2.4. Analysis of the Results

All main results from each tested fuel for different engine operation load have been summarized in Table 6. Dominant diameters in particle number are the smallest for 50% ATJ for every engine operation load. The same conclusion is reached concerning the range of dominant diameters in the particulate matter mass. To compare the cumulative values of the relative number and relative mass of particles for tested fuels, it was found that due to increasing the engine operation load, the diameter of 90% of the relative number of particles for each of the tested fuels decreases. The same happens when the content of the alternative fuel in tested fuels is higher, so the diameter of 90% of relative number of particles was the smallest for high load engine operation fueled 50% ATJ. The differences between the fuels and engine operation load in diameter of 90% of the relative number of particles shows last rows of Table 7.

Table 7. Main results from researches depending on tested fuel and engine operation load.

Engine Operation Load	Tested Fuel		
	Jet A-1	30% ATJ	50% ATJ
	Range of dominant diameters in particles number [nm]		
low	25.5–124.1	25.5–107.5	19.1–93.1
medium	25.5–107.5	22.1–93.1	16.5–80.6
high	16.5–107.5	22.1–93.1	16.5–80.6
	Range of dominant diameters in particulate matter mass [nm]		
low	25.5–220.7	34.0–165.6	25.5–165.6
medium	29.4–165.5	25.5–124.1	25.5–124.1
high	25.5–143.3	25.5–143.3	22.1–124.1
	The most dominant diameter [nm]		
low	60.4	52.3	34.0
medium	52.3	34.0	34.0
high	34.0	34.0	34.0
	90% of relative number particles is less than [nm]:		
low	100	80	70
medium	80	70	60
high	70	70	55

On the basis of the obtained data, the intensity of the number of particles E_{PN} was determined for the analyzed engine operation load ranges, using the measured numerical concentration of particles C_{PN} for a given engine load and the volumetric flow rate of exhaust gases for individual fuels. The intensity of the emission of particulate matter E_{PM} was also determined in the tested ranges of the engine operation load, based on the mass concentration of the particulate matter of the C_{PM} and the volumetric flow rate of exhaust gases for tested fuels. The intensity of the number and emission of particles was compared between the tested fuels and the analyzed engine load areas (Figure 8).

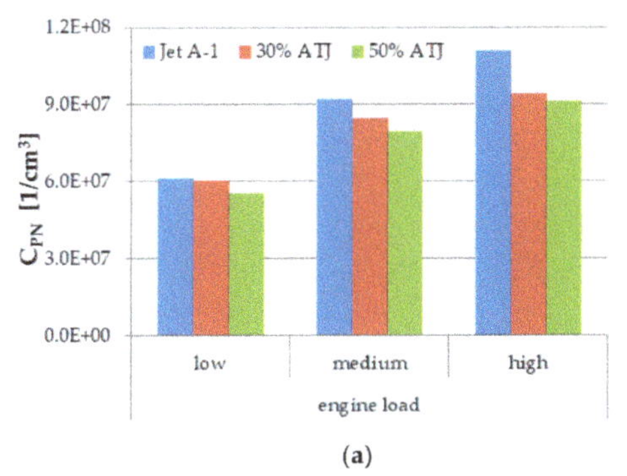

(a)

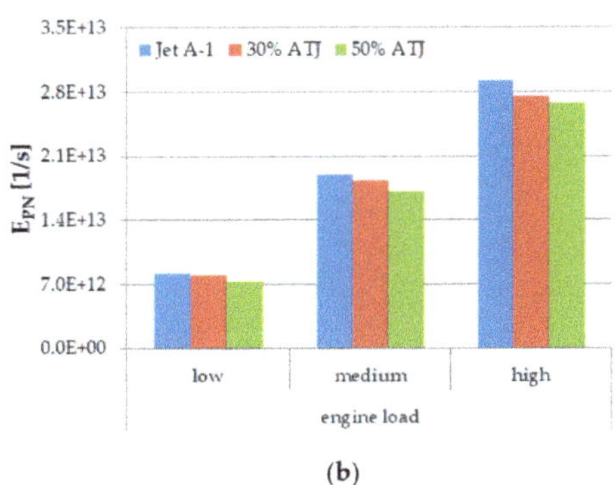

(b)

Figure 8. The number concentration C_{PN} (**a**) and intensity of the particles number emission E_{PN} (**b**) for tested fuels in three ranges of engine operation.

The number concentration of C_{PN} particles and the intensity of the number of particles E_{PN} increases with increasing engine load for all analyzed fuels. In all engine operating ranges, the particles number and number concentration are the highest for conventional Jet A-1 fuel and the lowest for fuel containing 50% of the alternative fuel.

The mass concentration of C_{PM} particulate matter decreased with increasing engine operation load in the case of fuels containing alternative fuel and was the highest at the medium engine load for Jet A-1 fuel (Figure 9). On the other hand, the intensity of E_{PM} particulate matter emission increased with increasing engine operation load for all tested fuels and was the highest for high engine load in the case of Jet A-1 fuel. The lowest particulate matter emission intensity was found for the fuel containing 50% ATJ fuel, which for particular engine operation load accounted for about 53% of the particulate emission intensity for Jet A-1 fuel.

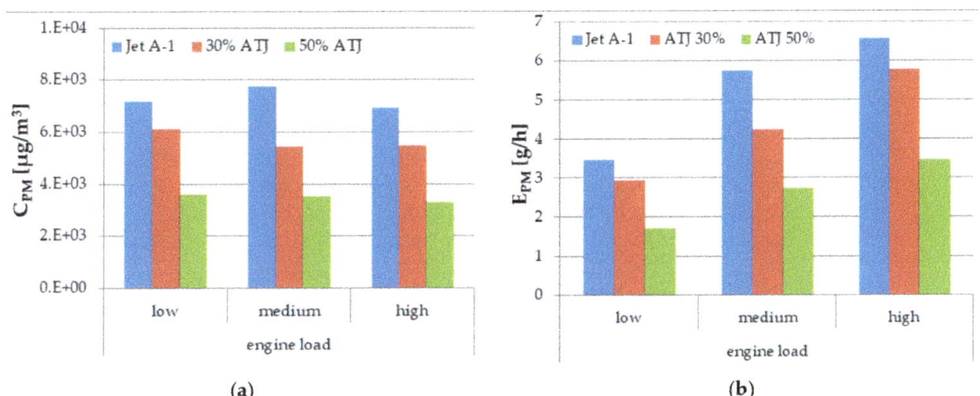

(a)

(b)

Figure 9. Mass concentration C_{PM} (**a**) and emission intensity of particulate matter E_{PM} (**b**) for tested fuels in three engine operation ranges.

The W_{PN} and W_{PM} coefficients were also determined, which determine the number and mass of particles, respectively, formed from one kilogram of fuel consumed by the engine. The values of the discussed coefficients are shown in Figure 10. Based on the charts below, it was found that the highest average number of particles is generated when the engine is fueled with conventional Jet A-1 fuel and at high engine operation load it amounts to 7.45×10^{15} units (Figure 10a). On the other hand, for high engine load the lowest average number of particles per kilogram of fuel used is for fuel containing 50% ATJ fuel and it amounts to 5.05×10^{15} units, thus it is about 32% lower than for Jet A-1. In the case of the particulate matter mass coefficient (Figure 10b), the highest mean value of the coefficient was equal to 0.53 for Jet A-1 fuel for medium engine loads, and the lowest for fuel containing 50% ATJ for high engine loads. The largest difference between the average value of the particulate matter mass factor, amounting to 63%, was found at the medium engine operation load between Jet A-1 fuel and the fuel containing 50% ATJ.

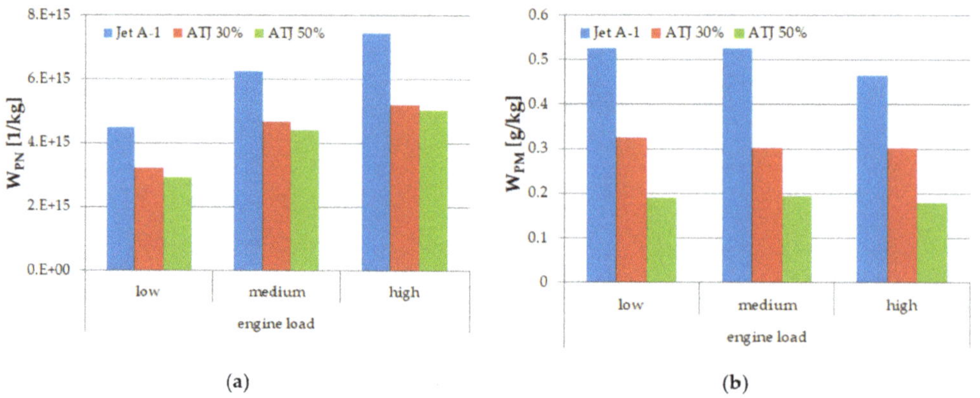

Figure 10. Particle number index W_{PN} (**a**) and particle mass index W_{PM} (**b**) for tested fuels in three engine load ranges.

6. Conclusions

Given the problems that the aviation industry is currently facing, the development of alternative fuels to power aircraft is inevitable. Over the past 10 years, over 200,000 flights have already been made using alternative fuels. Thanks to constant research and new solutions, the biofuels and sustainable fuels sector is constantly developing. In relation to the research carried out, using alternative fuel based on alcohol which is ATJ fuel, can have positive impact on the concentration of number and mass of particles compared to Jet A-1, but also negative impact on the emission of harmful gaseous compounds. It is crucial that the maximum engine load for a mixture of 30% ATJ and Jet A-1 and 50% ATJ and Jet A-1 was about 90% of maximum engine load for pure Jet A-1. Thus when comparing the emission of gaseous compound and particulate matter, attention should also be paid to the maximum achievable engine load for a given fuel mixture. For nitrogen oxides, hydrocarbons and carbon monoxide increasing the content of ATJ fuel in mixture of Jet A-1 and ATJ results in an increase of emission of these gaseous compounds in almost every engine operation load situation that was analyzed. As shown in the graphs of cumulative values of the relative particles number and relative mass of particulate matter for tested fuels, it was found that due to the increasing engine operation load, the diameter of 90% of the relative number of particles for each of the tested fuels decreases. The same happens when the content of the alternative fuel in tested fuels is higher, so the diameter of 90% of relative number of particles was the smallest for high load engine operation fueled by mixture of 50% ATJ and Jet A-1. Due to studies it is found out that the highest average number of particles is generated when the engine is fueled with conventional Jet A-1 fuel and the lowest average number of particles per kilogram of fuel used is for fuel containing

50% ATJ fuel. Thus, it can be concluded that the addition of ATJ has a positive effect on the number and mass concentration of particles.

The aim and the main conclusion of the above comprehensive analysis in the field of exhaust gas emissions from biofuels is that the emission of gaseous compounds is not necessarily lower than with the use of conventional Jet A-1 fuel, but in terms of the entire life cycle of biofuels, they are still less harmful to environment than conventional fuels. On the other hand, the emissions of particulate matter, in contrast to toxic compounds, is better when using a mixture of conventional fuel and biofuel than when using pure conventional fuel. Further recommendation is to maximize the proportion of biofuels, as far as possible and with all the safety and technical aspects of the engine, because they give measurable effects in the form of reduced particulate emissions.

Author Contributions: Conceptualization, R.J. and P.K.; methodology, R.J.; software, P.K.; validation, P.K. and R.J.; formal analysis, P.K.; investigation, P.K.; resources, R.J.; data curation, R.J.; writing—original draft preparation, P.K.; writing—review and editing, R.J.; visualization, P.K.; supervision, R.J.; project administration, R.J.; funding acquisition, R.J. Both authors have read and agreed to the published version of the manuscript.

Funding: The APC was funded by Interdisciplinary Rector's Grant, grant number ERP/33/32/SIGR/0004.

Institutional Review Board Statement: Not applicable.

Informed Consent Statement: Not applicable.

Conflicts of Interest: The authors declare no conflict of interest.

References

1. IRENA. *Biofuels for Aviation, Technology Brief*; International Renewable Energy Agency: Abu Dhabi, UAE, 2017.
2. Bosch, J.; Hoefnagels, R.; Jong, S.; Slade, R. *Aviation Biofuels: Strategically Important, Technically Achievable, Tough to Deliver*; Grantham Institute, Imperial College London: London, UK, 2017.
3. Airbus Global Market Forecast, Cities, Airports & Aircraft, 2019–2038. 2019. Available online: http://gmf.airbus.com/assets/pdf/Airbus_Global_Market_Forecast_2019-2038.pdf?v=1.0.1 (accessed on 14 January 2020).
4. European Aviation Environmental Report 2019, EASA, EEA, Eurocontrol. Available online: https://ec.europa.eu/transport/sites/transport/files/2019-aviation-environmental-report.pdf (accessed on 12 January 2020).
5. Merkisz, J.; Idzior, M.; Lijewski, P.; Fuć, P.; Karpiuk, W. *The Analysis of the Quality of Fuel Spraying in Relation to Selected Rapeseed Oil Fuels for the Common Rail System*; SAE International: Warrendale, PA, USA, 2008.
6. Braun-Unkhoff, M.; Riedel, U. Alternative fuels in aviation. *CEAS Aeronaut. J.* **2015**, *6*, 83–93. [CrossRef]
7. Hakes, J. *A Declaration of Energy Independence: How Freedom from Foreign Oil Can Improve National Security, Our Economy, and the Environment*; Wiley: Hoboken, NJ, USA, 2008.
8. Renewable Energy Directive, European Commission. Available online: https://ec.europa.eu/energy/topics/renewable-energy/renewable-energy-directive/overview_en (accessed on 10 March 2021).
9. European Parliament; Council of the European Union. Directive (EU) 2018/2001 of the European Parliament and of the Council of 11 December 2018 on the promotion of the use of energy from renewable sources (recast). *Off. J. Eur. Union* **2018**, *PE/48/2018/REV/1*, 82–209.
10. Atmanli, A.; Yilmaz, N. An experimental assessment on semi-low temperature combustion using waste oil biodiesel/C3-C5 alcohol blends in a diesel engine. *Fuel* **2019**, *260*, 116357. [CrossRef]
11. IATA. *Sustainable Aviation Fuels Fact Sheet*; International Air Transport Association: Montreal, QC, Canada, 2019.
12. Saha, S.; Sharma, A.; Purkayastha, S.; Pandey, K.; Dhingra, S. Bio-plastics and Biofuel: Is it the Way in Future Development for End Users? In *Plastics to Energy Fuel, Chemicals, and Sustainability Implications*; Plastics Design Library Series; Elsevier: Amsterdam, The Netherlands, 2019; pp. 365–376.
13. ATAG. *Beginner's Guide to Sustainable Aviation Fuel*; Air Transport Action Group: Geneva, Switzerland, 2017.
14. Carriquiry, M.A.; Du, X.; Timilsina, G.R. Second generation biofuels: Economics and policies. *Energy Policy* **2011**, *39*, 4222–4234. [CrossRef]
15. Oregon State University. *Generation of Biofuels*; Bioenergy Education Initiative, Oregon State University: Corvallis, OR, USA.
16. Kostova, B. *Current Status of Alternative Aviation Fuels*; U.S. Department of Energy: Washington, DC, USA, 2017.
17. Shonnard, D.R.; Williams, L.; Kalnes, T.N. Camelina-Derived Jet Fuel and Diesel: Sustainable Advanced Biofuels. *Environ. Prog. Sustain. Energy* **2010**, *29*, 382–392. [CrossRef]
18. Power-to-Liquids Potentials and Perspectives for the Future Supply of Renewable Aviation Fuel, German Enivronment Agency. 2016. Available online: https://www.umweltbundesamt.de/sites/default/files/medien/377/publikationen/161005_uba_hintergrund_ptl_barrierrefrei.pdf (accessed on 10 March 2021).

19. Schmidt, P.; Batteiger, V.; Roth, A.; Weindorf, W.; Raksha, T. Power-to-Liquids as Renewable Fuel Option for Aviation: A Review. *Chem. Ing. Tech.* **2018**, *90*, 127–140. [CrossRef]
20. RAND Corporation. *Infrastructure, Safety and Environment*; RAND Corporation: Santa Monica, CA, USA, 2009.
21. ICAO. *Sustainable Aviation Fuels Guide*; International Civil Aviation Organization: Montreal, QC, Canada, 2018.
22. ASTM. *ASTM Standardization News, D7566 Takes Flight*; ASTM International: West Conshohocken, PA, USA, 2011.
23. CAAFI Commercial Aviation Alternative Fuel Initiative, Fuel Qualification. Available online: http://www.caafi.org/focus_areas/fuel_qualification.html (accessed on 13 August 2020).
24. Karpiuk, W.; Borowczyk, T.; Idzior, M.; Smolec, R. The Evaluation of the Impact of Design and Operating Parameters of Common Rail System Fueled by Bio-Fuels on the Emission of Harmful Compounds. In Proceedings of the 2016 International Conference on Sustainable Energy, Environment and Information Engineering (SEEIE 2016), Bangkok, Thailand, 20–21 March 2016.
25. IATA. *Sustainable Aviation Fuels Roadmap*; International Air Transport Association: Montreal, QC, Canada, 2015.
26. Chevron Products Company, Aviation Fuels Technical Review. 2007. Available online: https://skybrary.aero/bookshelf/books/2478.pdf (accessed on 10 March 2021).
27. Biomass Magazine. PNNL Technology Clears Way for Ethanol-Derived Jet Fuel. 2018. Available online: http://biomassmagazine.com/articles/15369/pnnl-technology-clears-way-for-ethanol-derived-jet-fuel (accessed on 12 January 2020).
28. Csonka, S. *The State of Sustainable Aviation Fuel (SAF)*; CAAFI Webinar Series; International Civil Aviation Organization: Montreal, QC, Canada, 2020.
29. Zschocke, A.; Scheuermann, S.; Ortner, J. *High Biofuel Blends in Aviation (HBBA), ENER/C2/2012/420-1, Final Report*; Lufthansa: Cologne, Germany, 2012.
30. Johnston, G. *Alcohol to Jet—Isobutanol, ICAO Seminarium on Alternative Fuels 2017*; ICAO: Montreal, QC, Canada, 2017.
31. LanzaTech. No Carbon Left Behind: Alcohol-to-Jet. 2018. Available online: https://www.iata.org/contentassets/8dc7f9f4c38247ae8f007998295a37d5/jennifer-holmgren-vf-panel-session-2.pdf (accessed on 10 March 2021).
32. Jasiński, R. Evaluation of Nanoparticles Mass and Size Emissions from Aircraft Engines. Ph.D. Thesis, Poznan University of Technology, Poznań, Poland, 2019.
33. Merkisz, J.; Pielecha, J. The on-road exhaust emissions characteristics of SUV vehicles fitted with diesel engines. *Siln. Spalinowe* **2011**, *50*, 58–72.
34. Braun-Unkhoff, M.; Riedel, U.; Wahl, C. About the Emissions of Alternative Jet Fuels. *CEAS Aeronaut. J.* **2017**, *8*, 167–180. [CrossRef]
35. Riebl, S.; Braun-Unkhoff, M.; Riedel, U. A study on the emissions of alternative aviation fuels. *J. Gas Turbines Power* **2017**, *139*, 081503. [CrossRef]
36. Lobo, P.; Durdina, L.; Smallwood, G.J.; Rindlisbacher, T.; Siegerist, F.; Black, E.A.; Yu, Z.; Mensah, A.A.; Hagen, D.E.; Miake-Lye, R.C.; et al. Measurement of aircraft engine non-volatile PM emissions: Results of the Aviation-Particle Regulatory Instrumentation Demonstration Experiment (A-PRIDE) 4 Campaign. *Aerosol Sci. Technol.* **2015**, *49*, 472–484. [CrossRef]
37. Park, K.; Cao, F.; Kittelson, D.B.; McMurry, P.H. Relationship between particle mass and mobility for diesel exhaust particles. *Environ. Sci. Technol.* **2003**, *37*, 577–583. [CrossRef] [PubMed]
38. Petzold, A.; Marsh, R.; Johnson, M.; Miller, M.; Sevcenco, Y.; Delhaye, D.; Ibrahim, A.; Williams, P.; Bauer, H.; Crayford, A.; et al. Evaluation of methods for measuring particulate matter emissions from gas turbines. *Environ. Sci. Technol.* **2011**, *45*, 3562–3568. [CrossRef] [PubMed]

Article

Research of Energy and Ecological Indicators of a Compression Ignition Engine Fuelled with Diesel, Biodiesel (RME-Based) and Isopropanol Fuel Blends

Alfredas Rimkus [1,2], **Jonas Matijošius** [2,3,*] **and Sai Manoj Rayapureddy** [2]

[1] Department of Automobile Transport Engineering, Technical Faculty, Vilnius College of Technologies and Design, Olandų g. 16, LT-01100 Vilnius, Lithuania; a.rimkus@vtdko.lt

[2] Department of Automobile Engineering, Faculty of Transport Engineering, Vilnius Gediminas Technical University, J. Basanavičiaus g. 28, LT-03224 Vilnius, Lithuania; sai-manoj.rayapureddy@stud.vgtu.lt

[3] Institute of Mechanical Science, Vilnius Gediminas Technical University, Basanavičiaus g. 28, LT-03224 Vilnius, Lithuania

* Correspondence: jonas.matijosius@vgtu.lt; Tel.: +370-684-041-69

Received: 27 March 2020; Accepted: 6 May 2020; Published: 11 May 2020

Abstract: This article presents the results of a study of energy and ecological indicators at different engine loads (*BMEP*) adjusting the Start of Injection (*SOI*) of a Compression Ignition Engine fuelled with blends of diesel (D), rapeseed methyl ester (RME)-based biodiesel and isopropanol (P). Fuel blends mixed at D50RME45P5, D50RME40P10 and D50RME30P20 proportions were used. Alcohol-based fuels, such as isopropanol, were chosen because they can be made from different biomass-based feedstocks and used as additives with diesel fuel in diesel engines. Diesel fuel and its blend with 10% alcohol have almost the same thermal efficiency (*BTE*). In further examination of energy and ecological indicators, combustion parameters were analysed at *SOI* 6 CAD BTDC using *AVL BOOST* software (*BURN* subprogram). Increasing alcohol content in fuel blends led to a reduced cetane number, which prolonged the ignition delay phase and intensified heat release in the premixed combustion phase. Higher combustion temperatures and oxygen content in the fuel blends increased NO_x emissions. Lower C/H ratios and higher O_2 levels affected by RME and isopropanol reduced smoke emissions.

Keywords: compression ignition (CI) engine; biodiesel; isopropanol; combustion; energy indicators; ecological indicators

1. Introduction

The rising demand for fossil fuels and the environmental issues concerning their use are the biggest challenges that people face today. The transportation sector alone contributes up to 30% of the world's harmful emissions [1]. As the demand for fossil fuels continues to rise faster than its supply, fossil fuels deplete, which in turn drives the price of such fuels as diesel and petrol up [2–4]. The absence of a replacement for vehicles that run on liquid fuels, and a high price of electric vehicles made the automotive industry devote its resources to finding alternative fuels as a replacement and to decreasing the emissions concerned with environmental problems [5–7]. Recent studies revealed that greenhouse gases and harmful combustion chamber emissions can be significantly reduced by using fuels, such as alcohol and biodiesel, as primary alternatives [8–16].

In order to beat the fossil fuel deficiency and control the increasing demand of natural gas, opportunities for using alternative fuels in internal combustion engines have been searched [17]. The ease of handling and storing alcohol and biodiesel makes them promising substitutes for fossil fuels [9,18–21].

Modern biofuel production technologies are sufficiently developed and focused on the production of alcohols, which are very widely used as fuel additives [22]. Mostly ethanol is produced, the production technologies of which have been well known since ancient times. As a fuel additive ethanol has many disadvantages, the most important of which is its corrosivity [23]. Therefore, the selection of an alternative fuel blend for these studies focused on propanol, which is less corrosive and is close to petroleum diesel in its properties and calorific value [24]. In addition, when blended with diesel (10% v/v), propanol performs better in terms of emissions and noise than ethanol [25]. This choice was also based on its cheapness compared to other higher order alcohols like butanol and pentanol [26].

The use of alternative fuels in standard internal combustion engines most often requires modifying such engines. Our proposed three-component fuel mixture allows the replacement of conventional fuels (diesel) with alternative analogues, the use of biodiesel must take into account many operational aspects, such as e.g., *CFPP*, and alcohol propanol allows one to partially solve this problem. Therefore, the chosen three-component diesel-biodiesel-propanol blend allows us to solve the difficult task of using alternative fuels without engine modifications. The three-component mixtures mentioned are rarely found in the literature and are usually defined by a narrow analysis of fuel consumption and some exhaust parameters, and a more detailed analysis of heat release and other parameters is postponed for further research [27].

The performance of the diesel engine with the blend of *n*-propanol at different proportions like 10% by volume is attainable [28]. Diesel engines can use propanol-diesel blends containing 10 to 20% propanol without significantly affecting engine performance. It is further concluded that 10–20% of propanol-diesel blends are beneficial in reducing harmful fumes and NO_x emissions [29]. *n*-Propanol/diesel blends higher than 30% showed lower soot density due to the predominant effect of increasing spontaneous oxygen content in *n*-propanol/diesel mixtures [30].

It was observed that when the isopropanol concentration exceeds 15%, the combustion temperature and *BTE* performance starts to increase in biodiesel [31], especially in biodiesel-diesel-isopropanol blends (80%/10%/10%) [27]. Isopropanol improves the cold-flow limit in blends with both diesel and biodiesel, cooling it when operating at low temperatures [32]. 15% Isopropanol content in diesel allows for improved engine performance, lower smoke and NO_x emissions (at low to medium loads), but increases brake specific fuel consumption (*BSFC*) due to its lower calorific value [33]. Increasing the isopropanol content to 55% compared to diesel led to an increase in nitrogen oxide emissions (by an average of 139%), a reduction in carbon monoxide emissions (45%) and an increase in CO_2 emissions (by an average of 17%). However, no significant change in unburnt hydrocarbon emissions was observed [34].

Mixing different alcohols (ethanol-isopropanol and butanol) (*EPB*) with diesel at the ratio of 20 and 40% did not demonstrate a significant increase in heat release and combustion pressure compared to that of diesel. They have been found to have a lower molecular weight flux permeability than diesel, and the flame light index was lower with the use of the aforementioned additives in diesel, which contributed to lower soot emissions [35]. On the other hand, this led to a shorter initial combustion duration (*ICD*) and major combustion duration (*MCD*) conditions [11]. Changing injection strategies for *EPB* blends revealed that pilot injection reduces the heat release rate and the peak pressure, while dual injection improves fuel economy, reduces NO_x emissions, at the same time increasing soot [35].

The blend of 30% *EPB* alcohols with gasoline improves *BTE* and slightly increases CO (4.2%), hydrocarbon (HC) (18.9%) and NO_x emissions (5.5%) compared to mineral gasoline [36]. Correspondingly, adding 1% of water to *EPB* blends (10% *EPB* and 90% mineral gasoline) resulted in an even better ecological effect—a decrease in CO of up to 7.5% and in NO_x of up to 12.4%, respectively [12]. The emulsion blend of fuel and water reduces local areas where the maximum temperature is reached, resulting in decreased nitrogen activity and NO_x concentration [37]. However, using isopropanol additive in gasoline alone significantly increases HC emissions at low inlet air temperatures with increasing isopropanol concentration in isopropanol-gasoline blends [38].

When using 25% isopropanol in petrol blends with combustion control, there is a direct relationship between octane number and combustion parameters of isopropanol. Start of combustion (*SOC*) is delayed by increasing isopropanol content in the blend because isopropanol is more resistant to jerky engine operation [39]. By reducing spark time in a petrol engine, isopropanol-gasoline blends (with isopropanol content up to 30%) had lower NO_x emissions than those found at the initial spark time [40].

Oxygen content is the key parameter that differs in fossil fuels and biodiesel [41–43]. Environmentally friendly biodiesel produces clean and renewable energy, thus making it the alternative hope [44–46]. A study found that rapeseed methyl ester (RME) produced lower CO_2 emissions due to a lower carbon-to-hydrogen ratio as compared to diesel [47]. Methanol, ethanol are most commonly used in fuel blends with biodiesel, and currently the second ACB wave will make butanol cheaper [48]. The use of proponol in fuel blends with biodiesel is a rare case, determined by the greater development of other alcohol production technologies [20,49]. Lower heating values for B90Pr10 (90% biodiesel and 10% propanol fuel mixtures) and cetane number increased *BSFC* and fuel gas temperatures while reducing *BTE* [50]. Table 1 lists the properties of pure diesel, RME and isopropanol [51].

Table 1. Properties of 100% pure diesel, rapeseed methyl ester and isopropanol.

Properties	Diesel	Rapeseed Methyl Ester	Isopropanol
Density (kg/m^3)	843	877	785.1
Mass Fraction (*% mass*): Carbon	86.3	77.5	60
Hydrogen	13.7	12	13.4
Oxygen	0	10.5	26.6
Stoichiometric AFR	14.3	12.5	10.4
Lower Heating Value (*LHV*)(*MJ/kg*)	42.3	37.8	32.8
Cetane number	51	48	12
Auto-ignition temperature ($^\circ C$)	250	240	399

The aim of the research is to reveal the performance, combustion and emission characteristics of IC engines using pure diesel and fuel blends with different proportions of diesel, rapeseed methyl ester-based biodiesel and isopropanol.

2. Materials and Methods

2.1. Engine Testing Equipment

The engine used for testing is a 1.9 Turbocharged Direct Injection (TDI) diesel engine with a VP37 (BOSCH, Stuttgart, Germany) electronic controlled distribution type fuel pump. The start of the injection (*SOI*) was controlled by the Engine Electronic Control Unit (ECU) and it was a single injection strategy. Figure 1 presents a detailed image of the tested engine and its parts, while research conducted by other authors [52–55] and Table 2 lists the test engine parameters.

The brake torque M_B (Nm) was determined on a load bench with a measurement error of ±1.2 Nm. Hourly fuel consumption B_f (kg/h) was found using electronic scales *SK-5000* with a 0.5% measurement error. The pressure in the cylinder was measured using a piezoelectric sensor *GG2-1569* mounted on a glow plug with a sensitivity of 15.8 ± 0.09 pC/bar. Cylinder pressure values were recorded using the LabView Real software at an interval of 0.176 CAD. The pressure in the engine intake manifold was measured using an OHM HD 2304.0 pressure gauge (Delta, Padova, Italy) with a measurement error of ± 0.0002 MPa. The intake air and the exhaust gas temperature were measured using K-type thermocouples *IR 8839* accurate to ± 1.5 °C. The exhaust gas concentration was determined using a DiCom 4000 gas analyzer (AVL, Graz, Austria). CO_2 measurement accuracy was 0.1% vol., CO—0.01% vol., HC—1 ppm, NO_x—1 ppm, and smoke absorption coefficient—0.01 m^{-1}.

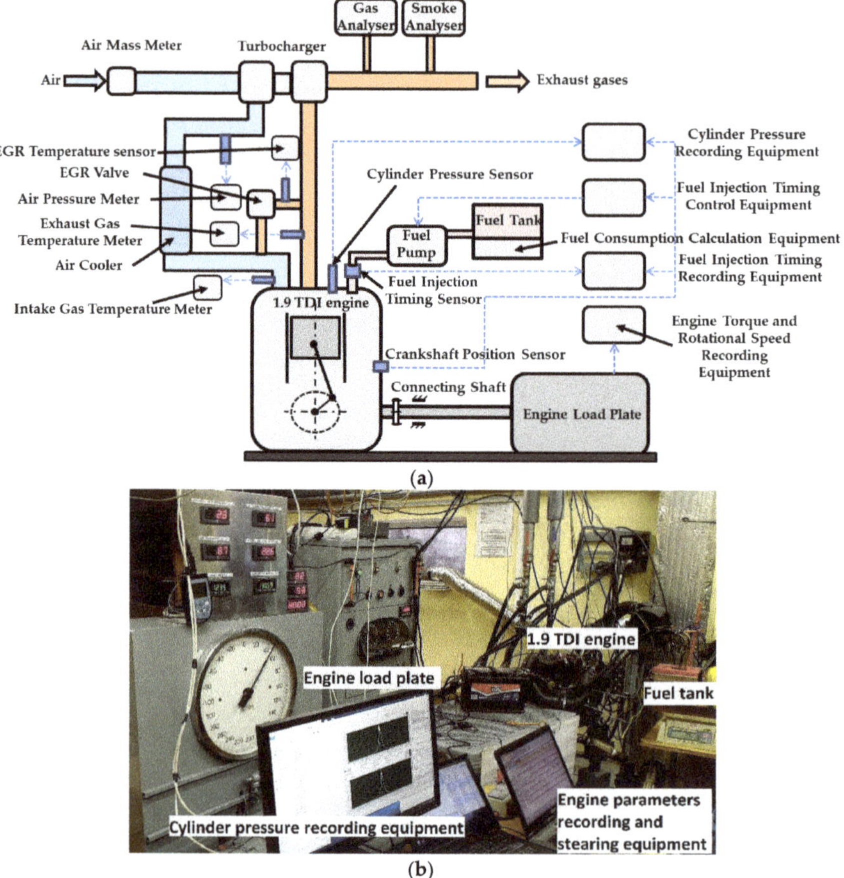

Figure 1. Image of the test engine: (**a**) test bench scheme; (**b**) test rig.

Table 2. Main parameters of the 1.9 TDI diesel engine.

Parameter	Value
Displacement (cm^3)	1896
No. of cylinders	4
Compression ratio	19.5
Power (kW)	66 (4000 rpm)
Torque (Nm)	180 (2000–25,000 rpm)
Bore (mm)	79.5
Stroke (mm)	95.5
Intake valve opening at	13 CAD before TDC
Intake valve closing at	25 CAD after BDC
Exhaust valve opening at	28 CAD before BDC
Exhaust valve closing at	19 CAD after TDC
Fuel injection	Direct injection (single)
Fuel injection-pump design	Axial-piston distributor injection pump
Nozzle type	Hole-type
Nozzle and holder assembly	Two-spring
Nozzle opening pressure (bar)	200

Statistical calculations of type A uncertainties were used to measure exhaust. Type A uncertainties were used to determine the standard deviation for repeated measurements, where $u(x)$ is uncertainty, n—repeatability of measurements, and $s(\overline{x})$—reliability [55,56]:

$$u(x) \;=\; s(\overline{x}) \;=\; \frac{s(x)}{\sqrt{n}} \tag{1}$$

where $\overline{x}$ is the mean repeated value; $s(x)$ is a standard deviation; $s(\overline{x})$ is a standard deviation of the mean. Uncertainty ranges u(x) of exhaust components are presented in Table 3.

Table 3. Uncertainty ranges $u(x)$ of exhaust components.

Exhaust Component	Number of Cycles	$\overline{x}$	Standard Uncertainly $u(x)$
CO (g/kWh)	4	830	0.0036
CO$_2$ (g/kWh)	4	895	0.0041
HC (g/kWh)	4	0.07	0.0005
NO$_x$ (g/kWh)	4	13.5	0.0079
Smoke (m^{-1})	4	7.3	0.0026

2.2. Fuels and Test Conditions

Tests were conducted using 100% pure diesel fuel and fuel blends prepared using different proportions of diesel (D), rapeseed methyl ester (RME)-based biodiesel and isopropanol (P). The first blend contained 50% diesel, 30% rapeseed methyl ester and 20% isopropanol (D50RME30P20), the second blend had 50% diesel, 40% rapeseed methyl ester and 10% isopropanol (D50RME40P10), and the third one 50% diesel, 45% rapeseed methyl ester and 5% isopropanol (D50RME45P5). The properties like density, mass fraction and lower heating value of the blends were calculated using the following formula:

$$\text{properties of fuel blends} = \sum \left[(\text{percentage of fuel blend} \times \text{property})\right] \tag{2}$$

Table 3 presents a comparison of the calculated properties of the fuel blends with the standard diesel fuel. Uncertainties were calculated according to the model B [55]. Standard uncertainties were calculated according to the formula:

$$u_c \;=\; \sqrt{u_c^2(f)} \tag{3}$$

where $u_c{}^2(f)$ is the total uncertainty dispersion.

The uncertainty calculations for each fuel blend are presented in Table 4.

Table 4. Comparison of fuel properties of different fuel blends used and uncertainty ranges u_c of each fuel blend parameter.

	D100	D50RME45P5	D50RME40P10	D50RME30P20
Density (kg/m^3)	843	855.4	850.8	844.4
u_c of Density (kg/m^3)	0.008	0.0037	0.0032	0.0026
Mass Fraction (%): Carbon	86.3	81.025	80.15	78.4
u_c of Carbon (%)	0.00333	0.00322	0.00215	0.00203
Hydrogen	13.7	12.92	12.99	13.13
u_c of Hydrogen	0.00064	0.00058	0.00047	0.00044
Oxygen	0	6.055	6.86	8.47
u_c of Oxygen	0	0.00008	0.00009	0.00012
Lower Heating Value (MJ/kg)	42.3	39.8	39.55	39.05
u_c of Lower Heating Value (MJ/kg)	0.00814	0.00726	0.00633	0.00589
Cetane Number (-)	51	34.49	31.84	27.94
u_c of Cetane Number (-)	0.0255	0.0344	0.0467	0.0592

Engine tests were carried out at the engine speed of $n = 2000$ rpm and engine brake torque M_B was 30, 60 and 90 Nm, which corresponds to the Brake Mean Effective Pressure (*BMEP*) 0.2 MPa, 0.4 MPa and 0.6 MPa in the first experimental tests step. These are the loads of a city car running of the $\approx$ 50 km/h, $\approx$ 80 km/h and $\approx$ 100 km/h speeds. During load-changing tests, fuel Start or Injection Timing (*SOI* $\approx$ 2 CAD BTDC) was controlled by the engine electronic control unit. There were (*BMEP* = 0.3 MPa) injection timing was adjusted (*SOI* = 0 ... 16 CAD BTDC) by modulating the *SOI* control signal in the second experimental test step. Injection timing was adjusted to determine the variation of engine performance using fuel mixtures of different properties under different combustion conditions. The Energy Indicators (Hourly Fuel Consumption B_f, Brake Specific Fuel Consumption *BSFC*, Brake Thermal Efficiency (*BTE*)) and Ecological Indicators (emission of carbon monoxide CO, carbon dioxide CO_2, nitrogen oxides NO_x, hydrocarbons CH, and smoke) were measured and calculated at different engine loads and by adjusting start of injection. The results of all the tested fuel blends will be analysed comparing them with results of diesel fuel. Experimental tests were performed to ensure repeatability of the experiments. Several experiment design parameters were singled out, such as fuel type used, engine rotation speed, engine load torque, and fuel injection angle.

The density difference between the fuels was found to decrease with increasing alcohol concentration as seen in Table 4. Since the molecular mass of alcohols is lower than that of diesel and biodiesel [57–60], with the alcohol content increasing from 5% to 10%, the density was found to decrease by 1.45%, and with an additional increase from 10% to 20%, the density decreased by nearly 0.17%. When compared to diesel, fuel blends with a 5 and 10% alcohol content tended to be denser, and the D50RME30P20 blend tended to have a lower density compared to conventional fuel as calculated and presented in Table 4.

A lower heat value was found to differ less with an increase in the alcohol percentage share as seen in Table 4. The lower heat value difference increased to 6.5% when increasing the fuel concentration from 5% to 10%, and a further change of the concentration from 10% to 20% led to the difference increasing by nearly 7.7%. The lower heat value highly depends on carbon and hydrogen content in fuel [61–63]. So, with a relatively higher carbon and hydrogen content in diesel compared to their content in other fuel blends, the lower heat value of conventional fuel was found to be higher than that of other fuel blends, which declined with increasing alcohol content as seen in Table 4.

The difference in the cetane number decreased with increasing alcohol content as seen in Table 4. The conventional diesel fuel is known for being rich in paraffin, which helps it achieve a higher cetane number compared to other fuel blends [64]. When increasing the alcohol content from 5% to 10%, the difference was found to increase to 37.5%, and a further increase in alcohol content to 20% led to an increase in the difference of nearly 45%. As seen in Table 4, the cetane number steadily decreased with alcohol content compared to that of the conventional fuel. The fuel stability was ensured by producing fuel blends right before testing and feeding them to the engine.

2.3. Tools for Numerical Analysis of the Combustion Process

Due to a significant change in the cetane number in diesel and other prepared fuel blends, analysing changes in performance of the engine by calculating combustion characteristics and comparing them with those of diesel is necessary. *AVL BOOST* software was used in calculating the combustion characteristics with the help of *BURN* subprogram. *BURN* analysis was conducted having created a digital model of the 1.9 TDI diesel engine used in the experiment as seen in Figure 2. The digital model of the engine was constructed by selecting the required elements from a displayed element catalogue in *AVL BOOST*. The analysis of combustion requires general data on engine parameters, fuel and data describing the operating point.

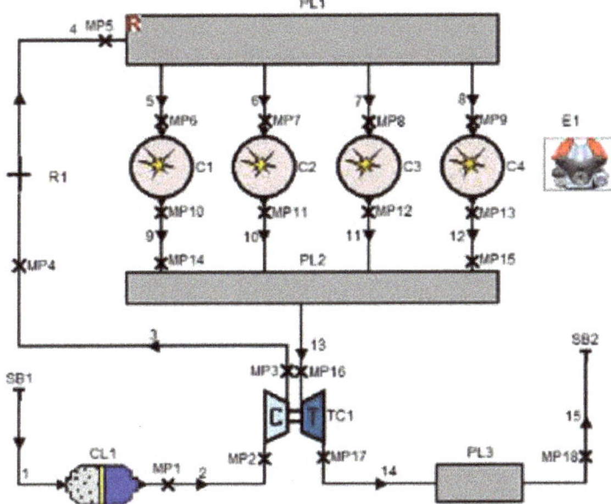

Figure 2. Digital Engine Model used in AVL BOOST.

Engine parameters and experiment results were uploaded in the BURN subprogram and run to get the combustion characteristics. Changes in the in-cylinder pressure and temperature, heat release rate (*ROHR*) and the mass fraction burnt (*MFB*) were calculated in this research:

$$ROHR = \frac{dx}{d\alpha} = \frac{6.908}{\alpha_{CD}} \cdot (m_v + 1) \cdot \left(\frac{\alpha - \alpha_{SOC}}{\alpha_{CD}}\right)^{m_v} \cdot e^{-6.908 \cdot \left(\frac{\alpha - \alpha_{SOC}}{\alpha_{CD}}\right)^{(m_v+1)}} \tag{4}$$

$$dx = \frac{dQ}{Q} \tag{5}$$

$$MFB = \int_{\alpha_{SOC}}^{\alpha} \frac{dQ}{d\alpha \cdot Q(\alpha)} \cdot d\alpha = 1 - e^{-6.908 \cdot \left(\frac{\alpha - \alpha_{SOC}}{\alpha_{CD}}\right)^{(m_v+1)}}, \ \alpha > \alpha_{SOC} \tag{6}$$

where Q—total fuel heat input; α—crank angle; m_v—combustion shape parameter; α_{SOC}—start of combustion; α_{CD}—combustion duration.

3. Results and Discussion

The experimental tests were carried out in two steps: (a) changing the engine load *BMEP* (0.2; 0.4 and 0.6 MPa); (b) changing *SOI* (0 . . . 16 CAD *BTDC*), and *BMEP* = 0.3 MPa. After plotting the graphs from the (b) experiment results, a polynomial curve was drawn with a degree of 2 for all the energy and ecological parameters to get a change trend.

3.1. Energy Indicators

Brake Specific Fuel consumption (BSFC) of diesel fuel was low at all loads compared to other fuel blends, as observed in Figure 3a. When increasing alcohol content, fuel consumption tended to increase with D50RME30P20 being at the maximum. Having replaced 50% of diesel by a blend of biodiesel and propanol and increased the concentration of propanol up to 20% (D50RME30P20), BSFC increased ~9% due to a 7.7% reduction in LHV (Table 4) and a change in combustion process. With BMEP = 0.3 MPa, the analysis of BSFC from the perspective of injection timing revealed a higher consumption of all the fuel blends (7–10%), and with advancing angle, the value tended to decrease and further increase after 8–12 CAD BTDC as seen in Figure 3b.

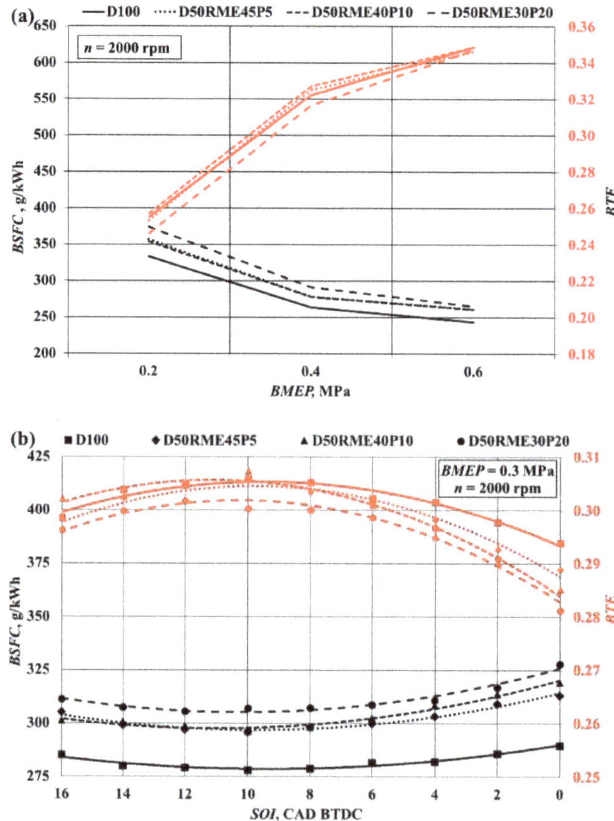

Figure 3. Dependence of Brake Specific Fuel Consumption and Brake Thermal Efficiency on different:
(**a**) loads; (**b**) injection timing.

At 8 CAD BTDC, D100 fuel was at its lowest on average. The lowest consumption of D-RME-P blends was achieved with injection timing of 10...12 CAD BTDC, as the cetane number of this fuel decreased to 23.06.

BTE of D50RME45P5 and D50RME40P10 fuel blends at *BMEP* 0.2 and 0.4 MPa was close to that of diesel, and *BTE* of D50RME30P20 was 3.5% (at *BMEP* 0.2 MPa) to 1.8% (at *BMEP* 0.4 MPa) lower as seen in Figure 3a. With a *BMEP* = 0.6 MPa, *BTE* of all fuels was slightly different and reached up to 0.35%.

Observing the dependence curve of efficiency on injection timing in Figure 3b, at *BMEP* = 0.3 MPa, *BTE* of the D50RME30P20 fuel blend was ~1.8% lower than that of diesel. *BTE* of fuel blends with 5% and 10% isopropanol content was lower (2.5 ... 1.2%) at *SOI* = 0 ... 6 CAD BTDC, but at *SOI* = 8 ... 12 CAD BTDC, *BTE* of D50RME45P5 and D50RME40P10 was the same as *BTE* of D100.

3.2. Ecological Indicators

Carbon dioxide comparative emissions (g/kWh) were found to decrease for all the fuels with increasing load as seen in Figure 4a as *BTE* increased and *BSFC* decreased. At a low load (*BMEP* = 0.2 MPa), CO_2 emissions of the blends containing isopropanol were 0.8 ... 1.9% higher compared to diesel but increasing the load to *BMEP* = 0.6 MPa resulted in ~0.8% lower CO_2 emissions for the D50RME40P10 fuel blend.

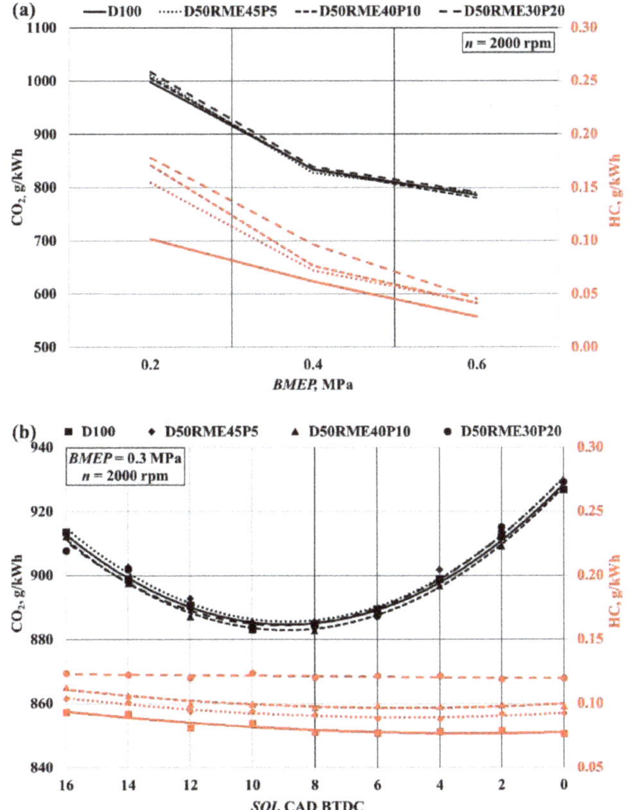

Figure 4. Dependence of carbon dioxide emissions and hydrocarbons on different: (**a**) loads; (**b**) injection timing.

This was due to a 2.1% decrease in the C/H ratio (Table 4) and the fact that at higher loads the *BTE* of all fuels was similar. Checking the dependence of emissions on injection timing revealed that all the fuels showed a gradual decrease in emissions with advancing injection timing as seen in Figure 4b, *BTE* increased, and smoke emissions decreased (Figure 5b). Diesel was found to have similar CO_2 emissions compared to emissions of other fuel blends. CO_2 emissions of the D50RME40P10 fuel blend at various *SOIs* were only ~0.2% lower than emissions of D100. Even though RME and C/H ratio of isopropanol is lower, increased fuel consumption increases CO_2 emissions.

Hydrocarbon emissions of all the fuels were found to decrease with increasing load as seen in Figure 4a, as the combustion temperature increased [65,66]. Having the lowest emissions as compared to all the other fuels, diesel also tended to have a steady decrease pattern of ~38%, ~45% and ~60% (compared to the D50RME45P5, D50RME40P10 and D50RME30P20 fuel blends in hydrocarbon emissions). When increasing the alcohol content, fuel blends tended to have higher hydrocarbon emissions due to increasing alcohol base of the fuel, however, HC emissions were low compared to petrol engine [52]. The observation of the dependence of hydrocarbon emissions on injection timing revealed that with an increase in alcohol content, emissions tended to increase as seen in Figure 4b. HC emissions of D50RME45P5, D50RME40P10 and D50RME30P20 increased by ~15%, ~28% and ~58% compared to D100 at *SOI* ≈ 6 CAD BTDC. With the engine running on all the fuels, hydrocarbon emissions showed a slight growth trend when injection timing was advanced more than 6 CAD BTDC.

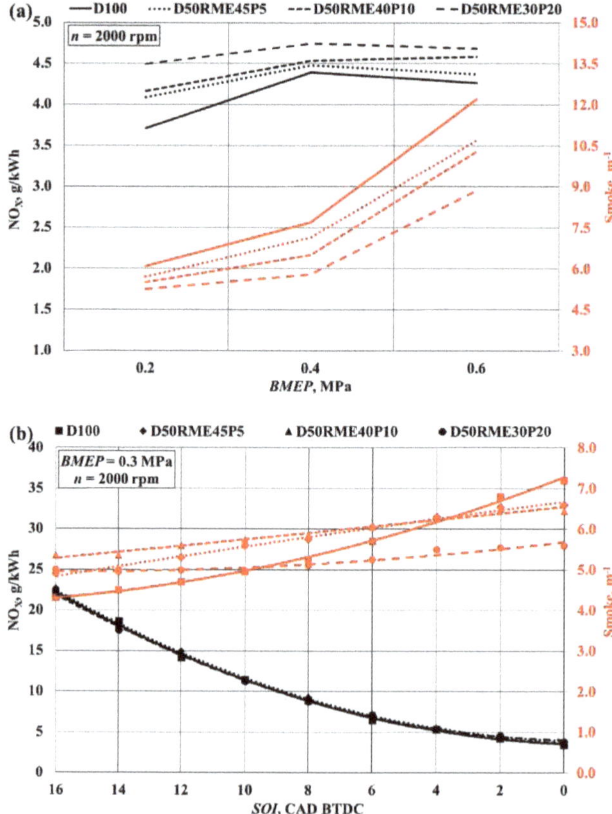

Figure 5. Dependence of nitrogen oxide emissions and smoke on different: (**a**) loads; (**b**) injection timing.

Nitrogen oxide emissions for D50RME45P5, D50RME40P10, D50RME30P20 at a low load (*BMEP* = 0.2 MPa) were ~10%, ~12%, and ~21% higher compared to those of D100 fuel as seen in Figure 5a. This was mainly due to the increased ignition delay due to a low cetane number of isopropanol (see the Dependence of the rate of heat release and the mass burnt fraction on the crank angle degree figure bellow.) and the increased oxygen concentration in the fuel blend (Table 4).

With an increase in the load (*BMEP* = 0.6 MPa), the difference in NO_x emissions was reduced to ~4%, ~7% and ~10% as the effect of ignition delay on different fuels was reduced. While analysing the dependence of nitrogen oxide emissions on the injection timing, emissions were found to rise steadily with increasing injection timing as seen in Figure 5b, because combustion occurred at a lower volume and higher temperatures [67]. All the fuels were found to follow a similar pattern, but after a more careful analysis, diesel emissions were found to be lower (1 … 4%) compared to emissions of other fuel blends. An earlier injection timing reduced the difference between NO_x emissions of diesel and fuel blends [68–70].

The smoke level of fuels tended to increase with increasing load as seen in Figure 5a, as fuel mass increases per cycle and the air-to-fuel ratio decreases. Increasing the isopropanol concentration in fuel blends reduces smoke level, and this effect is more intense with an increasing engine load [71]. At a low load (*BMEP* = 0.2 MPa), the D50RME45P5, D50RME40P10 and D50RME30P20 fuel smoke emissions decreased by ~6%, ~10% and ~14%, respectively, in comparison to pure diesel. Higher oxygen concentrations and lower C/H ratios resulted in lower D50RME30P20 smoke emissions. Increasing the load to *BMEP* = 0.6 MPa increased the smoke reduction effect to ~12%, ~16%.

The analysis of the dependence of smoke levels on injection timing revealed that the levels tended to decrease with advancing the injection timing as seen in Figure 5b. Interestingly, throughout the *SOI* study range (0...16 CAD BTDC), the combustion performance resulted in the largest reduction in diesel smoke emissions from ~7.25 m^{-1} to ~4.25 m^{-1} (~40%), while the D50RME30P20 smoke emissions decreased from ~5.7 to ~5 m^{-1} (~12%). Therefore, at *SOI* = 0...8 CAD BTDC (low advanced injection timing), D50RME30P20 had the lowest smoke emissions, and at *SOI* = 10...16 CAD BTDC (high advanced injection timing), smoke emissions of D100 fuel were the lowest.

3.3. Combustion Characteristics

The analysis of combustion characteristics was conducted with the engine operating at *BMEP* = 0.3 MPa (*n* = 2000 *rpm*). Changing *SOI* (0 ... 16 CAD BTDC) allowed comparing energy and ecological performance of the engine running on different fuels (Figures 3b, 4b and 5b) and concluding that with ignition timing being 6 CAD BTDC engine efficiency is close to the maximum, and smoke and NO$_x$ emissions are relatively low. The pressure values of all the fuels at *SOI* = 6 CAD BTDC obtained during the experiment as seen in Figure 6 were uploaded in the AVL BOOST (BURN subprogram) to get the combustion characteristics. Since the result at *SOI* = 6 CAD BTDC was relatively good, the combustion characteristics of fuel blends, such as in-cylinder pressure, pressure rise, rate of heat release (*ROHR*) and mass fraction burned (*MFB*), were analysed and compared to those of pure diesel at that particular degree of ignition timing.

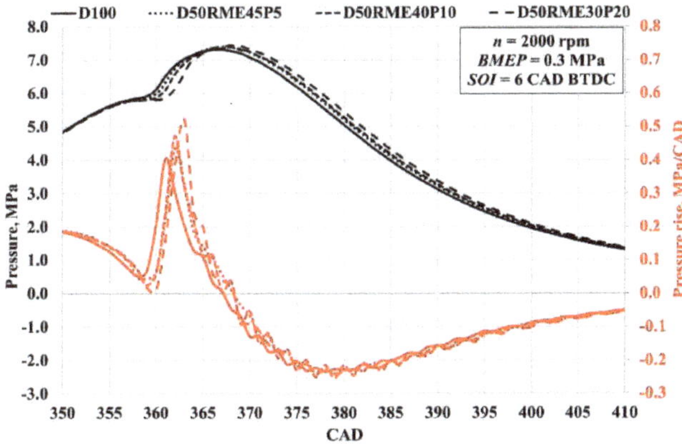

Figure 6. Dependence of pressure and pressure rise in the cylinder on the crank angle degree.

Conventional diesel fuel (D100) showed the maximum pressure of ~7.32 MPa at 367 CAD, while the maximum pressures of D50RME45P5, D50RME40P10 and D50RME30P20 were ~1.0%, ~1.1% and ~1.7% higher as seen in Figure 6. There was also a delayed burning with isopropanol.

A combustion-driven pressure starts to increase the earliest with the engine running on diesel. The maximum pressure rise of ~0.40 MPa/CAD was observed at 361 CAD as seen in Figure 6. The maximum pressure rise increases with an increase in alcohol percentage. ~7.5% (361.5 CAD), ~17% (362 CAD) and ~29% (363 CAD) was the maximum increase for D50RME45P5, D50RME40P10 and D50RME30P20.

The maximum rate of heat release of diesel fuel was lower at 32.0 J/deg than that of other fuel blends as seen in Figure 7. The maximum rate of 34.8 J/deg (~8% higher) was observed with the D50RME4P5 fuel blend, while the maximum rate of D50RME40P10 was 38.1 J/deg (~19% higher) and that of D50RME30P20—46.4 J/deg (~44% higher). Isopropanol reduces the cetane number (Table 4), prolongs the ignition delay phase and significantly increases the maximum *ROHR* during the premixed combustion phase [58]. The *LHV* of isopropanol is lower, but this is offset by the higher fuel content

(Figure 7), and the diffusion combustion phase produces a similar amount of heat for all fuels. Higher maximum temperatures during premixed combustion phase increase the formation of nitrogen oxides, but allows for a better combustion of soot at the end of the combustion process [69–71].

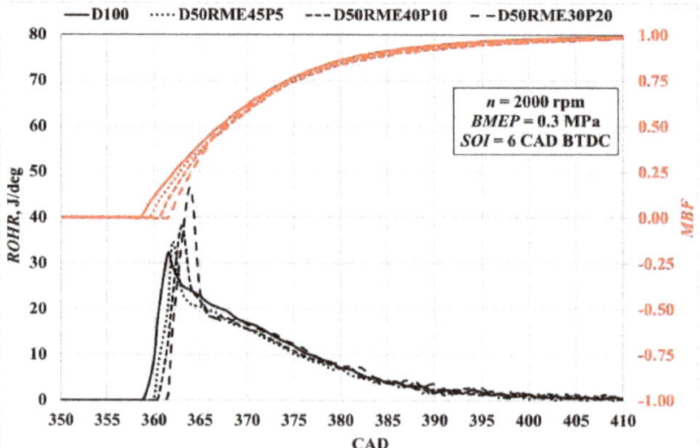

Figure 7. Dependence of the rate of heat release and the mass burnt fraction on the crank angle degree.

The mass burn fraction diagram in Figure 7 confirms the prolongation of the ignition delay phase in fuel blends with a higher alcohol content. Ignition delay for D100 was ~5 CAD, D50RME45P5—~6 CAD, D50RME40P10—~7 CAD and D50RME30P20—~8 CAD.

Although ignition delay is longer for blends with isopropanol [72], oxygen concentration in blends significantly increases due to RME and isopropanol. This significantly accelerates the combustion process during the premixed combustion phase, and 0.5 *MBF* of all fuels is available at ~7.5 CAD ATDC. The diffusion combustion phase *MBF* intensity is similar for all fuels, although the fuel consumption of D50RME30P20 increased by ~9% (Figure 3), which was offset by increased injection rate due to a lower isopropanol viscosity and faster combustion driven by a higher oxygen concentration. 99% of the D100 fuel mass ends up burning ~410 CAD ATDC, D50RME45P5—~412 CAD ATDC, D50RME40P10—~413 CAD ATDC and D50RME30P20—~414 CAD ATDC.

4. Conclusions

The analysis of the energy, ecological and combustion parameters of diesel 100, D50RME45P5, D50RME40P10 and D50RME30P20 in a turbocharged direct injection diesel engine at the speed (*n*) of 2000 rpm and under various loads and injection timings allows making the following conclusions:

(1)   RME and isopropanol reduce *LHV* and the cetane number of fuel blends, but increase the oxygen concentration in the blend and lower the C/H ratio.

(2)   D50RME30P20 brake specific fuel consumption increased by ~9% compared to D100 and *BTE* decreased by ~1.8% due to a 7.7% reduction in *LHV* and a change in the combustion process. The maximum *BTE* of the D50RME40P10 fuel blend was equal to D100 efficiency having advanced the injection timing of the fuel blend ~2 CAD. This offset the increase in the ignition delay due to a low propanol cetane number (~12).

(3)   Carbon dioxide emissions of all fuels are similar, but the best carbon dioxide effect was obtained with the D50RME40P10 fuel blend. At medium loads, CO_2 emissions of this blend declined by ~0.2% compared to diesel, though fuel consumption increased by ~6%, as the C/H ratio of the fuel blend was 2.1% lower. A more advanced injection timing (*SOI* = 8 … 10 CAD BTDC) at the minimum fuel consumption allows achieving lower CO_2 emissions.

(4)   Isopropanol has a greater impact on nitrogen oxide emissions at low loads. NO_x emissions of D50RME45P5, D50RME40P10 and D50RME30P20 increased by ~10%, ~12% and ~21% due to a

higher oxygen concentration in the blends and a higher combustion temperature. With increasing load, an increase in NO_x emissions was lower (~4%, ~7% and ~10%) as a low isopropanol cetane number had a lesser effect on the ignition delay phase and the heat release rate during the premixed combustion phase. With an early injection timing, NO_x emissions increased, but the impact of alcohol was lower.

(5)   Having replaced diesel with fuel blends at a low load resulted in a ~6%, ~10% and ~14% reduction in smoke and an average load reduction of ~12%, ~16% and ~28%. Smoke was reduced by lower C/H ratios and increased oxygen content in the fuels. As the load increased, the *BTE* of the fuel blends increased more intensively, which further reduced smoke emissions. In the case of early injection timing ($SOI = 8 \ldots 16$ CAD BTDC), smoke emissions of the fuel blends changed (decreased) less intensively due to changed fuel characteristics compared to pure diesel.

(6)   At a low engine load ($BMEP = 0.3$ MPa), the average rotation speed ($n = 2000$ rpm), the fixed injection timing ($SOI = 6$ CAD BTDC) and the replacement of diesel by fuel blends with a higher alcohol content (5%, 10% and 20%) resulted in ignition delay changing from ~5 CAD to ~6 CAD, ~7 CAD and ~8 CAD. A greater ignition delay (accumulates more fuel) and a higher oxygen content in fuel during the premixed combustion phase increased the heat release intensity by ~8%, ~19% and ~44%, which in turn increased the pressure rise by ~8%, ~17% and ~29% in the thermodynamic load of the crank mechanism. During the diffusion combustion phase, the combustion heat release of all the fuels examined was similar.

(7)   The authors plan to continue research of these three-component blends by increasing the share of alternative fuels in the blends and assessing the impact of the EGR system when using these blends.

Author Contributions: Author Contributions: Conceptualization, A.R. and S.M.R.; methodology, A.R., S.M.R. and J.M.; software, S.M.R. and A.R.; formal analysis, A.R. and J.M.; validation, A.R. and S.M.R.; writing—original draft preparation, A.R. and J.M.; writing—review and editing, A.R. and J.M.; supervision, A.R., and J.M.; project administration, J.M. All authors have read and agreed to the published version of the manuscript.

Funding: This research received no external funding.

Acknowledgments: Part of the combustion characteristics reported in this article were obtained using the engine numerical analysis tool AVL BOOST, acquired by signing a Cooperation Agreement between AVL Advanced Simulation Technologies and the faculty of the Transport Engineering of Vilnius Gediminas Technical University.

Conflicts of Interest: The authors declare no conflict of interest.

Abbreviations and Nomenclature

ACB	Acetoneb—utanol producing process
ATDC	After Top Dead Centre (CAD)
AVL	Anstalt für Verbrennungskraftmaschinen List
BDC	Bottom Dead Center
B_f	Fuel mass consumption (kg/h)
BMEP	Brake Mean Effective Pressure (MPa)
BSFC	Brake Specific Fuel Consumption (g/kWh)
BTDC	Before Top Dead Center (CAD)
BTE	Brake Thermal Efficiency
CA	Crank Angle (degree)
CFPP	Cold filter plugging point
CO	Carbon monoxide
CO_2	Carbon dioxide
CN	Cetane Number
CV	Calorific Value
D	Diesel fuel

ECU	Electronic Control Unit
EPB	Ethanol-Propanol and butanol fuel blend
HC	Hydrocarbons
IC	Internal combustion
ICD	Initial combustion duration (CAD)
LHV	Lower Heating Value (MJ/kg)
M_B	Brake torque (Nm)
MFB	Mass fraction burned
MCD	Major combustion duration (CAD)
n	Rotational speed of the crankshaft (rpm)
NO_x	Nitrogen Oxide
O_2	Oxygen
P	Isopropanol
ROHR	Rate of heat release (J/deg)
RME	Rapeseed Methyl Ester
SOI	Start of Injection (CAD)
TDC	Top Dead Center
TDI	Turbocharged Direct Injection

References

1. Enriquez, A.; Benoit, L.; Dalkmann, H.; Brannigan, C. GIZ Sourcebook 5e Transport and Climate Change. *Tech. Rep.* **2014**. [CrossRef]
2. Abas, N.; Kalair, A.; Khan, N. Review of fossil fuels and future energy technologies. *Futures* **2015**, *69*, 31–49. [CrossRef]
3. David, M. The role of organized publics in articulating the exnovation of fossil-fuel technologies for intra- and intergenerational energy justice in energy transitions. *Appl. Energy* **2018**, *228*, 339–350. [CrossRef]
4. Jia, T.; Dai, Y.; Wang, R. Refining energy sources in winemaking industry by using solar energy as alternatives for fossil fuels: A review and perspective. *Renew. Sustain. Energy Rev.* **2018**, *88*, 278–296. [CrossRef]
5. Tamilselvan, P.; Nallusamy, N.; Rajkumar, S. A comprehensive review on performance, combustion and emission characteristics of biodiesel fuelled diesel engines. *Renew. Sustain. Energy Rev.* **2017**, *79*, 1134–1159. [CrossRef]
6. Sładkowski, A. (Ed.) *Ecology in Transport: Problems and Solutions*; Lecture Notes in Networks and Systems; Springer International Publishing: Cham, Switzerland, 2020; Volume 124, ISBN 978-3-030-42322-3.
7. Bureika, G.; Steišūnas, S. Complex Evaluation of Electric Rail Transport Implementation in Vilnius City. *Transp. Probl.* **2016**, *11*, 49–60. [CrossRef]
8. Bhasker, J.P.; Porpatham, E. Effects of compression ratio and hydrogen addition on lean combustion characteristics and emission formation in a Compressed Natural Gas fuelled spark ignition engine. *Fuel* **2017**, *208*, 260–270. [CrossRef]
9. Chen, H.; He, J.; Zhong, X. Engine combustion and emission fuelled with natural gas: A review. *J. Energy Inst.* **2019**, *92*, 1123–1136. [CrossRef]
10. Ghadikolaei, M.A.; Cheung, C.S.; Yung, K.-F. Study of combustion, performance and emissions of diesel engine fueled with diesel/biodiesel/alcohol blends having the same oxygen concentration. *Energy* **2018**, *157*, 258–269. [CrossRef]
11. Li, Y.; Chen, Y.; Wu, G.; Lee, C.-F.F.; Liu, J. Experimental comparison of acetone-n-butanol-ethanol (ABE) and isopropanol-n-butanol-ethanol (IBE) as fuel candidate in spark-ignition engine. *Appl. Therm. Eng.* **2018**, *133*, 179–187. [CrossRef]
12. Li, Y.; Chen, Y.; Wu, G.; Liu, J. Experimental evaluation of water-containing isopropanol-n-butanol-ethanol and gasoline blend as a fuel candidate in spark-ignition engine. *Appl. Energy* **2018**, *219*, 42–52. [CrossRef]
13. Li, G.; Liu, Z.; Lee, T.H.; Lee, C.-F.F.; Zhang, C. Effects of dilute gas on combustion and emission characteristics of a common-rail diesel engine fueled with isopropanol-butanol-ethanol and diesel blends. *Energy Convers. Manag.* **2018**, *165*, 373–381. [CrossRef]
14. Li, M.; Xu, J.; Xie, H.; Wang, Y. Transport biofuels technological paradigm based conversion approaches towards a bio-electric energy framework. *Energy Convers. Manag.* **2018**, *172*, 554–566. [CrossRef]

15. Moula, M.E.; Nyári, J.; Bartel, A. Public acceptance of biofuels in the transport sector in Finland. *Int. J. Sustain. Built Environ.* **2017**, *6*, 434–441. [CrossRef]

16. Yan, X.; Gao, P.; Zhao, G.; Shi, L.; Xu, J.-B.; Zhao, T. Transport of highly concentrated fuel in direct methanol fuel cells. *Appl. Therm. Eng.* **2017**, *126*, 290–295. [CrossRef]

17. Zefreh, M.M.; Török, Á. Theoretical Comparison of the Effects of Different Traffic Conditions on Urban Road Traffic Noise. *J. Adv. Transp.* **2018**, *2018*, 7949574. [CrossRef]

18. Hoseinpour, M.; Sadrnia, H.; Tabasizadeh, M.; Ghobadian, B. Evaluation of the effect of gasoline fumigation on performance and emission characteristics of a diesel engine fueled with B20 using an experimental investigation and TOPSIS method. *Fuel* **2018**, *223*, 277–285. [CrossRef]

19. Jamuwa, D.; Sharma, D.; Soni, S. Experimental investigation of performance, exhaust emission and combustion parameters of stationary compression ignition engine using ethanol fumigation in dual fuel mode. *Energy Convers. Manag.* **2016**, *115*, 221–231. [CrossRef]

20. Yusri, I.; Mamat, R.; Najafi, G.; Razman, A.; Awad, O.; Azmi, W.H.; Ishak, W.; Shaiful, A. Alcohol based automotive fuels from first four alcohol family in compression and spark ignition engine: A review on engine performance and exhaust emissions. *Renew. Sustain. Energy Rev.* **2017**, *77*, 169–181. [CrossRef]

21. Zaharin, M.; Abdullah, N.; Najafi, G.; Sharudin, H.; Yusaf, T. Effects of physicochemical properties of biodiesel fuel blends with alcohol on diesel engine performance and exhaust emissions: A review. *Renew. Sustain. Energy Rev.* **2017**, *79*, 475–493. [CrossRef]

22. Szabados, G.; Bereczky, A. Experimental investigation of physicochemical properties of diesel, biodiesel and TBK-biodiesel fuels and combustion and emission analysis in CI internal combustion engine. *Renew. Energy* **2018**, *121*, 568–578. [CrossRef]

23. Žaglinskis, J.; Lukács, K.; Bereczky, A. Comparison of properties of a compression ignition engine operating on diesel–biodiesel blend with methanol additive. *Fuel* **2016**, *170*, 245–253. [CrossRef]

24. Matijošius, J.; Sokolovskij, E. Research into the Quality of fuels and their Biocomponents. *Transport* **2009**, *24*, 212–217. [CrossRef]

25. Pinzi, S.; Redel-Macías, M.; Candia, D.E.L.; Soriano, J.A.; Dorado, M. Influence of ethanol/diesel fuel and propanol/diesel fuel blends over exhaust and noise emissions. *Energy Procedia* **2017**, *142*, 849–854. [CrossRef]

26. Bereczky, A. The Past, Present and Future of the Training of Internal Combustion Engines at the Department of Energy Engineering of BME. In *Vehicle and Automotive Engineering*; Jármai, K., Bolló, B., Eds.; Springer Science and Business Media LLC: Cham, Switzerland, 2017; Volume 13, pp. 225–234.

27. Bencheikh, K.; Atabani, A.; Shobana, S.; Mohammed, M.; Uğuz, G.; Arpa, O.; Kumar, G.; Ayanoğlu, A.; Bokhari, A. Fuels properties, characterizations and engine and emission performance analyses of ternary waste cooking oil biodiesel–diesel–propanol blends. *Sustain. Energy Technol. Assess.* **2019**, *35*, 321–334. [CrossRef]

28. Yogesh, P.; Dinesh Sakthi, S.; Aravinth, C.; Sathiyakeerthy, K.; Susenther, M. Performance Test on Diesel-Biodiesel–Propanol Blended Fuels in CI Engine. *IJISRT* **2018**, *3*, 1–6.

29. Muthaiyan, P.; Gomathinayagam, S. Combustion Characteristics of a Diesel Engine Using Propanol Diesel Fuel Blends. *J. Inst. Eng. (India) Ser. C* **2016**, *97*, 323–329. [CrossRef]

30. Thillainayagam, M.; Venkatesan, K.; Dipak, R.; Subramani, S.; Sethuramasamyraja, B.; Babu, R.K. Diesel reformulation using bio-derived propanol to control toxic emissions from a light-duty agricultural diesel engine. *Environ. Sci. Pollut. Res.* **2017**, *24*, 16725–16734. [CrossRef]

31. Campos-Fernández, J.; Arnal, J.M.; Gomez, J.; LaCalle, N.; Dorado, M.P. Performance tests of a diesel engine fueled with pentanol/diesel fuel blends. *Fuel* **2013**, *107*, 866–872. [CrossRef]

32. Atmanli, A.; Atmanlı, A. Comparative analyses of diesel–waste oil biodiesel and propanol, n-butanol or 1-pentanol blends in a diesel engine. *Fuel* **2016**, *176*, 209–215. [CrossRef]

33. Şen, M. The effect of the injection pressure on single cylinder diesel engine fueled with propanol–diesel blend. *Fuel* **2019**, *254*, 115617. [CrossRef]

34. Jamrozik, A.; Tutak, W.; Pyrc, M.; Gruca, M.; Kocisko, M. Study on co-combustion of diesel fuel with oxygenated alcohols in a compression ignition dual-fuel engine. *Fuel* **2018**, *221*, 329–345. [CrossRef]

35. Lee, T.H.; Hansen, A.C.; Li, G.; Lee, T. Effects of isopropanol-butanol-ethanol and diesel fuel blends on combustion characteristics in a constant volume chamber. *Fuel* **2019**, *254*, 115613. [CrossRef]

36. Li, G.; Lee, T.H.; Liu, Z.; Lee, C.-F.F.; Zhang, C. Effects of injection strategies on combustion and emission characteristics of a common-rail diesel engine fueled with isopropanol-butanol-ethanol and diesel blends. *Renew. Energy* **2019**, *130*, 677–686. [CrossRef]
37. Chybowski, L.; Laskowski, R.; Gawdzińska, K. An overview of systems supplying water into the combustion chamber of diesel engines to decrease the amount of nitrogen oxides in exhaust gas. *J. Mar. Sci. Technol.* **2015**, *20*, 393–405. [CrossRef]
38. Uyumaz, A. An experimental investigation into combustion and performance characteristics of an HCCI gasoline engine fueled with n-heptane, isopropanol and n-butanol fuel blends at different inlet air temperatures. *Energy Convers. Manag.* **2015**, *98*, 199–207. [CrossRef]
39. Calam, A.; Aydoğan, B.; Halis, S. The comparison of combustion, engine performance and emission characteristics of ethanol, methanol, fusel oil, butanol, isopropanol and naphtha with n-heptane blends on HCCI engine. *Fuel* **2020**, *266*, 117071. [CrossRef]
40. Sivasubramanian, H.; Pochareddy, Y.K.; Dhamodaran, G.; Esakkimuthu, G.S. Performance, emission and combustion characteristics of a branched higher mass, C 3 alcohol (isopropanol) blends fuelled medium duty MPFI SI engine. *Eng. Sci. Technol. Int. J.* **2017**, *20*, 528–535. [CrossRef]
41. Baloch, H.A.; Nizamuddin, S.; Siddiqui, M.; Riaz, S.; Jatoi, A.S.; Dumbre, D.K.; Mubarak, N.; Srinivasan, M.; Griffin, G. Recent advances in production and upgrading of bio-oil from biomass: A critical overview. *J. Environ. Chem. Eng.* **2018**, *6*, 5101–5118. [CrossRef]
42. Dominković, D.F.; Bačeković, I.; Pedersen, A.S.; Krajačić, G. The future of transportation in sustainable energy systems: Opportunities and barriers in a clean energy transition. *Renew. Sustain. Energy Rev.* **2018**, *82*, 1823–1838. [CrossRef]
43. Trumbo, J.L.; Tonn, B.E. Biofuels: A sustainable choice for the United States' energy future? *Technol. Forecast. Soc. Chang.* **2016**, *104*, 147–161. [CrossRef]
44. Abbas, S.Z.; Kousar, A.; Razzaq, S.; Saeed, A.; Alam, M.; Mahmood, A. Energy management in South Asia. *Energy Strat. Rev.* **2018**, *21*, 25–34. [CrossRef]
45. Patel, P.D.; Lakdawala, A.; Chourasia, S.K.; Patel, R. Bio fuels for compression ignition engine: A review on engine performance, emission and life cycle analysis. *Renew. Sustain. Energy Rev.* **2016**, *65*, 24–43. [CrossRef]
46. Thangavelu, S.K.; Ahmed, A.S.; Ani, F.N. Review on bioethanol as alternative fuel for spark ignition engines. *Renew. Sustain. Energy Rev.* **2016**, *56*, 820–835. [CrossRef]
47. Imran, S.; Emberson, D.; Wen, D.; Diez, A.; Crookes, R.; Korakianitis, T. Performance and specific emissions contours of a diesel and RME fueled compression-ignition engine throughout its operating speed and power range. *Appl. Energy* **2013**, *111*, 771–777. [CrossRef]
48. Luo, H.; Zheng, P.; Bilal, M.; Xie, F.; Zeng, Q.; Zhu, C.; Yang, R.; Wang, Z. Efficient bio-butanol production from lignocellulosic waste by elucidating the mechanisms of Clostridium acetobutylicum response to phenolic inhibitors. *Sci. Total. Environ.* **2020**, *710*, 136399. [CrossRef] [PubMed]
49. Çelebi, Y.; Aydın, H. An overview on the light alcohol fuels in diesel engines. *Fuel* **2019**, *236*, 890–911. [CrossRef]
50. Yilmaz, N.; Ileri, E.; Atmanli, A. Performance of biodiesel/higher alcohols blends in a diesel engine. *Int. J. Energy Res.* **2016**, *40*, 1134–1143. [CrossRef]
51. Markov, V.A.; Gaivoronsky, A.I.; Grechov, L.V.; Ivashchenko, N.A. *Work of diesel engines on alternative fuels (aoa a ao oa)-In Russian*; Legion-Avtodata: Moscow, Russia, 2008; ISBN 978-5-88850-361-4.
52. Rimkus, A.; Žaglinskis, J.; Stravinskas, S.; Rapalis, P.; Matijošius, J.; Bereczky, A. Research on the Combustion, Energy and Emission Parameters of Various Concentration Blends of Hydrotreated Vegetable Oil Biofuel and Diesel Fuel in a Compression-Ignition Engine. *Energies* **2019**, *12*, 2978. [CrossRef]
53. Rimkus, A.; Pukalskas, S.; Matijošius, J.; Sokolovskij, E. Betterment of ecological parameters of a diesel engine using Brown's gas. *J. Environ. Eng. Landsc. Manag.* **2012**, *21*, 133–140. [CrossRef]
54. Fuć, P.; Lijewski, P.; Ziolkowski, A.; Dobrzyński, M. Dynamic Test Bed Analysis of Gas Energy Balance for a Diesel Exhaust System Fit with a Thermoelectric Generator. *J. Electron. Mater.* **2017**, *46*, 3145–3155. [CrossRef]
55. JCGM 100 2008. Evaluation of measurement data—Guide to the expression of uncertainty in measurement. Available online: https://www.bipm.org/utils/common/documents/jcgm/JCGM_100_2008_E.pdf (accessed on 13 April 2020).

56. Evaluation of measurement data—Supplement 1 to the "Guide to the expression of uncertainty in measurement"—Propagation of distributions using a Monte Carlo method. Available online: https://www.bipm.org/utils/common/documents/jcgm/JCGM_101_2008_E.pdf (accessed on 27 April 2020).

57. Gutarevych, Y.; Mateichyk, V.; Matijošius, J.; Rimkus, A.; Gritsuk, I.; Syrota, O.; Shuba, Y. Improving Fuel Economy of Spark Ignition Engines Applying the Combined Method of Power Regulation. *Energies* **2020**, *13*, 1076. [CrossRef]

58. Zoldy, M. Fuel Properties of Butanol–Hydrogenated Vegetable Oil Blends as a Diesel Extender Option for Internal Combustion Engines. *Period. Polytech. Chem. Eng.* **2019**, *64*, 205–212. [CrossRef]

59. Zöldy, M. Investigation of Correlation Between Diesel Fuel Cold Operability and Standardized Cold Flow Properties. *Period. Polytech. Transp. Eng.* **2019**. [CrossRef]

60. Kumar, S.; Pandey, A.K. Current Developments in Biotechnology and Bioengineering and Waste Treatment Processes for Energy Generation. *Curr. Dev. Biotechnol. Bioeng.* **2019**, 1–9. [CrossRef]

61. Zöldy, M. Improving Heavy Duty Vehicles Fuel Consumption with Density and Friction Modifier. *Int. J. Automot. Technol.* **2019**, *20*, 971–978. [CrossRef]

62. Ghosh, P.; Jaffe, S.B. Detailed Composition-Based Model for Predicting the Cetane Number of Diesel Fuels. *Ind. Eng. Chem. Res.* **2006**, *45*, 346–351. [CrossRef]

63. Gutarevych, Y.; Shuba, Y.; Matijošius, J.; Karev, S.; Sokolovskij, E.; Rimkus, A. Intensification of the combustion process in a gasoline engine by adding a hydrogen-containing gas. *Int. J. Hydrog. Energy* **2018**, *43*, 16334–16343. [CrossRef]

64. Kukharonak, H.; Ivashko, V.; Pukalskas, S.; Rimkus, A.; Matijošius, J. Operation of a Spark-ignition Engine on Mixtures of Petrol and N-butanol. *Procedia Eng.* **2017**, *187*, 588–598. [CrossRef]

65. Zöldy, M.; Hollo, A.; Thernesz, A. Butanol as a Diesel Extender Option for Internal Combustion Engines. *SAE Tech. Paper Ser.* **2010**. [CrossRef]

66. Hunicz, J.; Matijošius, J.; Rimkus, A.; Kilikevičius, A.; Kordos, P.; Mikulski, M. Efficient hydrotreated vegetable oil combustion under partially premixed conditions with heavy exhaust gas recirculation. *Fuel* **2020**, *268*, 117350. [CrossRef]

67. Caban, J.; Droździel, P.; Ignaciuk, P.; Kordos, P. THE IMPACT OF CHANGING THE FUEL DOSE ON CHOSEN PARAMETERS OF THE DIESEL ENGINE START-UP PROCESS. *Transp. Probl.* **2019**, *14*, 51–62. [CrossRef]

68. Kilikevičienė, K.; Kačianauskas, R.; Kilikevičius, A.; Maknickas, A.; Matijošius, J.; Rimkus, A.; Vainorius, D. Experimental investigation of acoustic agglomeration of diesel engine exhaust particles using new created acoustic chamber. *Powder Technol.* **2020**, *360*, 421–429. [CrossRef]

69. Rimkus, A.; Matijošius, J.; Bogdevicius, M.; Bereczky, A.; Török, Á. An investigation of the efficiency of using O_2 and H_2 (hydrooxile gas -HHO) gas additives in a ci engine operating on diesel fuel and biodiesel. *Energy* **2018**, *152*, 640–651. [CrossRef]

70. Zavadskas, E.K.; Čereška, A.; Matijošius, J.; Rimkus, A.; Baušys, R. Internal Combustion Engine Analysis of Energy Ecological Parameters by Neutrosophic MULTIMOORA and SWARA Methods. *Energies* **2019**, *12*, 1415. [CrossRef]

71. Zöldy, M. Potential future renewable fuel challanges for internal combustion engine. *Járművek és Mobilgépek II. évf* **2009**, *2*, 397–403.

72. Lijewski, P.; Fuć, P.; Dobrzynski, M.; Markiewicz, F. Exhaust emissions from small engines in handheld devices. *MATEC Web Conf.* **2017**, *118*, 16. [CrossRef]

 *energies*

Article

Resource Recovery from Waste Coffee Grounds Using Ultrasonic-Assisted Technology for Bioenergy Production

M. Mofijur [1],*, F. Kusumo [1,2], I. M. Rizwanul Fattah [1], H. M. Mahmudul [3], M. G. Rasul [3], A. H. Shamsuddin [2] and T. M. I. Mahlia [1]

[1]   School of Information, Systems and Modelling, Faculty of Engineering and Information Technology, University of Technology Sydney, Ultimo, NSW 2007, Australia

[2]   Institute of Sustainable Energy, Universiti Tenaga Nasional, Kajang 43000, Selangor, Malaysia

[3]   School of Engineering and Technology, Central Queensland University, Rockhampton, QLD 4701, Australia

*   Correspondence: MdMofijur.Rahman@uts.edu.au

Received: 12 March 2020; Accepted: 3 April 2020; Published: 7 April 2020

Abstract: Biodiesel is a proven alternative fuel that can serve as a substitute for petroleum diesel due to its renewability, non-toxicity, sulphur-free nature and superior lubricity. Waste-based non-edible oils are studied as potential biodiesel feedstocks owing to the focus on the valorisation of waste products. Instead of being treated as municipal waste, waste coffee grounds (WCG) can be utilised for oil extraction, thereby recovering an energy source in the form of biodiesel. This study evaluates oil extraction from WCG using ultrasonic and Soxhlet techniques, followed by biodiesel conversion using an ultrasonic-assisted transesterification process. It was found that n-hexane was the most effective solvent for the oil extraction process and ultrasonic-assisted technology offers a 13.5% higher yield compared to the conventional Soxhlet extraction process. Solid-to-solvent ratio and extraction time of the oil extraction process from the dried waste coffee grounds (DWCG) after the brewing process was optimised using the response surface methodology (RSM). The results showed that predicted yield of 17.75 wt. % of coffee oil can be obtained using 1:30 w/v of the mass ratio of DWCG-ton-hexane and 34 min of extraction time when 32% amplitude was used. The model was verified by the experiment where 17.23 wt. % yield of coffee oil was achieved when the extraction process was carried out under optimal conditions. The infrared absorption spectrum analysis of WCG oil determined suitable functional groups for biodiesel conversion which was further treated using an ultrasonic-assisted transesterification process to successfully convert to biodiesel.

Keywords: waste coffee grounds; ultrasonic-assisted technology; biodiesel; optimisation

---

1. Introduction

The limited reserves and increasing price of fossil fuels, as well as the adverse impact of fossil fuel combustion on climate change, has motivated researchers to find an alternative source of energy, e.g., renewable fuel sources [1–3]. Biodiesel is a renewable fuel consisting of a mixture of mono-alkyl esters and long-chain fatty acids, and is non-toxic, biodegradable and can be used in diesel engines with minimal modification [4–6]. Biodiesel is produced from different sources including edible oils, non-edible oils, and animal fats [7–11]. Currently, biodiesel production costs are calculated to be 4.4 times the cost of petroleum-derived diesel production [12]. Given that the current commercial biodiesel feedstocks are of edible oil origins such as palm and soy, government intervention in the biodiesel market will complicate the effects on food security. The use of WCG oil as biodiesel feedstock promotes biodiesel production from an alternative resource while reducing the issue of landfilling with food waste, which is prominent globally.

Coffee is the world's second-largest traded liquid commodity, after oil, with approximately 8 million tons of coffee produced each year [13]. A large amount of heat energy is used to convert green coffee beans into brown roasted beans in the brewing process, generating large amounts of volatile organic compounds (VOCs) [14]. The enormous demand for this beverage also produces a large quantity of residual waste after brewing. Every 1 ton of coffee beans produces 650 kg of coffee residue after brewing, known as WCG [15]. The global coffee industry produced an estimate of 9.34 million tons of waste in 2017, which was either incinerated, dumped in landfills or composted [16]. Every year Australia produces an estimated 75,000 tonnes of used ground coffee waste, and 93% of cafes send their WCGs to landfill. The annual domestic coffee consumption in Australia has reached almost 1.9 million 60 kg bags. On average, Australians consumed around 1.92 kilograms of coffee per person in 2017 [17]. The grounds that are used to make coffee are used only once and then immediately discarded. With rising rates of consumption, waste residues from the coffee industry (by-products from harvesting, processing, roasting and brewing stages of coffee production and processing) represent a challenge to worldwide directives aiming to reduce landfill volume. The inherent toxicity of several constituents within coffee also presents an environmental contamination concern [18]. Additionally, waste coffee grounds contribute towards the huge financial cost on taxpayers for running and maintaining landfills. Therefore, a combined solution of collection and reuse of WCG for alternate energy production would be beneficial to the coffee industry. WCG contains 15%–20% of lipid depending on the extraction technologies, which can be used as a source of bioenergy.

Production of oil from non-edible sources such as WCG can also help overcome the food versus fuel dispute [19]. The abundance of WCGs would also make it a readily available feedstock with a significantly lower production cost than edible oils [20]. WCG after oil extraction has also been identified as a suitable material for the production of garden fertiliser, feedstock for ethanol production, biogas production and fuel pellets [21]. However, the use of WCG oil for biodiesel production is still relatively new and requires further research before commercialisation.

This study aims to recover oil from WCG through an ultrasonic-assisted process and to convert it into biodiesel to reduce the volume of waste to landfill; reducing the greenhouse gas emissions associated with coffee waste in landfills. The WCG was subjected to ultrasonic-assisted oil extraction and compared with the Soxhlet extraction process before transesterification was used to produce WCG biodiesel by ultrasonication. The effect of extraction solvent, solvent to WCG ratio, ultrasonic power and ultrasonication period were studied to optimise oil extraction from the WCG. Further optimisation of parameters such as methanol to oil molar ratio, catalyst concentration, ultrasonic power and reaction time were done using Response Surface Methodology (RSM) to obtain the highest ester yield. RSM correlates the relationship between different response variables [22]. The effect of independent variables is determined by RSM which also creates a mathematical model which can be used for evaluating other relevant variables. The physicochemical properties of the obtained biodiesel were measured and compared with ASTM D6751 and EN 14214 biodiesel standards to determine its successful conversion to biodiesel [23–25].

2. Materials and Methods

WCG was obtained from a local store which was then oven-dried at 60° C. High purity analytical grade chemicals (Sigma–Aldrich) and solvents including 2-propanol (purity 99.7%), n-hexane (purity 99%), methanol (purity 99.8%), and potassium hydroxide were used to extract oil and convert into biodiesel.

2.1. Moisture

The moisture content of WCG was determined by measuring the mass before and after the drying of the collected samples. Drying was conducted in an electric oven (Tech-lab, stainless steel forced-air convection oven FAC-138SS) at 100° C for 36 h. Mass of WCG was weighed before drying and at intervals of 12 h. It was found that the mass stopped changing after 36 h which indicates the sample had dried. The following equation was used to determine the moisture content.

$$\text{Moisture content } (\%) = \frac{\Delta m}{m_i} \times 100\% \tag{1}$$

where Δm (g) is the changes between the final and initial mass of the sample, m_i (g) is the initial mass of WCG.

The moisture content of WCG was found to be 20%. For calculating more accurate oil yields, the mass of WCG was only weighed after the drying process. Table 1 shows the change of mass of WCG as drying time increases. Mass of WCG remained constant after the 24 h mark.

Table 1. Mass change of waste coffee grounds (WCG) with drying time.

Drying time (hours)	Mass (g)
0	156
12	130
24	125
36	125
48	125

2.2. Experimental Setup

The reactor used for Soxhlet extraction was equipped with a reflux condenser, one Soxhlet extractor, and a heating mantle. The temperature of the condenser was controlled using a refrigerator cooling bath WiseCircu®(Model: WCR-P8, Daihan Scientific, Gang-Won–Do, South Korea). The experimental setup of WCG Soxhlet extraction is illustrated in Figure 1a.

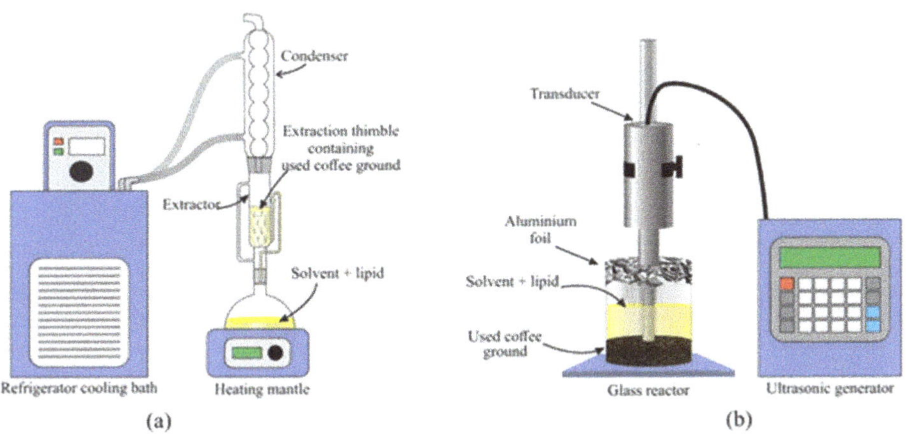

Figure 1. Schematic diagram of (a) Soxhlet extraction set-up, (b) ultrasonic-assisted extraction setup.

The equipment used for ultrasound extraction in this study was a Qsonica Q500-20 sonicator with a 1″ diameter tip (500 W power rating, 20 kHz frequency) ultrasonic probe. The sample was placed in 250 ml beaker made of borosilicate glass as a reactor, where the tip of the probe was fully immersed in the solvent and sample mixture. The probe was placed in the center of the reactor to ensure even ultrasonication of the entire sample. The precaution was taken to ensure that the probe tip was fully immersed to ensure direct sonication of the sample. Figure 1b shows a schematic of the ultrasonic-assisted extraction setup. The ultrasonication time, amplitude and frequency of ultrasonic waves can be changed using the sonicator system. However, the system was set at a moderate level to avoid energy wastage, deterioration of the sample, and to reduce risk of breaking of the equipment. The ultrasonication can also be set to continuous or pulsed modes. However, to increase the efficiency of the system, a pulsed mode was selected [26].

2.3. Soxhlet Extraction Method

In this step, 20 g of WCG was weighed in a cellulosic thimble before it was placed in a Soxhlet oil extractor. The oil was extracted using 300 ml of three different types of organic solvents including n-hexane, chloroform and methanol. Among the selected solvents, n-hexane and chloroform are polar in nature whereas methanol is a non-polar solvent. The latent heat of vaporisation of chloroform is the lowest followed by the n-hexane and methanol solvents. These organic solvents increase the yield of oil extraction [27] compared to other green solvents. The average cycle time was recorded as 15 min. To maintain this extraction cycle time, the temperature was varied based on the chosen solvent. The solvent-oil mixture was placed in a rotary evaporator (IKA RV 10 digital V rotary evaporator with vacuum) at 60 °C to separate the extracted oil. The extraction process was done three times for each solvent type and the average oil extraction yield was calculated by measuring the mass of the dried sample.

2.4. Ultrasonic-assisted Oil Extraction Method

In this step, 10 g of WCG was poured into the 500 ml beaker which also works as a reactor. Selected solvents including n-hexane, chloroform or methanol were added at a ratio 1:20 g/ml into ultrasonicator. The ultrasonic probe was immersed into a sample such that the tip was completely submerged in the solvent mixture. The ultrasonic waves were applied for 5 s with a stop interval of 2 s. Extraction time (25, 37.6, and 50 min) and ultrasonic amplitude (20%, 30%, and 40%) were the other parameters selected for optimisation. The temperature was not selected for optimisation because of continuous compression and rarefaction of the ultrasonic cavitation cycle, which produces heat within the mixture. Filter papers were used for gravitation filtration of the solvent mixture from WCG oil. The rotary evaporator was used to evaporate the sample at 60 °C. The oil yield was calculated using Equation (2).

$$\text{Oil yield percentage} = \frac{\text{mass of flask after exploration} - \text{mass of the empty flask}}{\text{mass of dried WCG}} \times 100\% \quad (2)$$

The molar mass of WCG oil was collected from literature as 862.8 g/mol [28].

2.5. Response Surface Methodology (RSM)

Design-Expert software version 11 (Stat-Ease Inc., Minneapolis, USA) was used to analyse and optimise experimental data. Analysis of variance (ANOVA) and RSM features of the software were utilised. To optimise the parameters of the oil extraction process, Box–Behnken experimental design was applied. The operating parameters such as the amplitude (X1), reaction time (X2) and n-hexane (X3), were varied to optimise oil yield (Y). The coded and uncoded levels of the Box–Behnken independent variables were presented in Table 2. The experimental data were analysed in the form of a mathematical model as follows:

$$Y = C_0 + \sum_{i=1}^{k} C_i\, X_i + \sum_{i=1}^{k} C_{ii}\, X_i^2 + \sum_{j=i+1}^{k} \cdot \sum_{i=1}^{k} C_{ij}\, X_{ij} \quad (3)$$

where, Y predicted the yield of WCG crude oil; Xi is the input independent parameter, C_0 and Ci are the intercept and regression first-order coefficient of the model, regression coefficient among ith and jth input parameters, and the number of input parameters is represented by k respectively. Cii is the regression quadratic coefficient of the model for the ith factor. Cij is the linear coefficient of the model for the interaction between the *i*th and *j*th factor.

Table 2. Independent input process variables used for the optimisation of biodiesel yield.

Input Process Variables	Units	Coded Factors	Coded Process Variables Levels		
			−1 Level	Center	+1 Level
Amplitude	%	X_1	20	30	40
Reaction time	min	X_2	25	37.5	50
n-hexane	%	X_3	15	22.5	30

2.6. Two-Step Esterification and Transesterification

The methyl ester was produced using a two-step esterification and trans-esterification process. The esterification process was carried out due to extracted oil having a very high acid value. In this process, the WCG crude oil sample was transferred into a reactor and then mixed with methanol at a molar ratio of 6:1 (methanol to oil). A 1 vol. % H_2SO_4 catalyst was added into the pre-heated oil (60 °C). An ultrasonication amplitude of 30% was applied for 60 to 105 min. The ultrasonic waves were applied for 5 s with 2-s rest intervals. Following the esterification process, the sample was transferred into a separating funnel. Two distinguishable liquid layers were observed, the top layer consisted of catalyst residues and methanol whereas the bottom layer contained esterified WCG oil.

The esterified oil was placed in a reactor. The required amount of catalyst KOH was weighed into a beaker along with methanol for mixing. Before pouring into the reactor, the mixture was heated and stirred up until the KOH pellets were completely dissolved. The ultrasonic probe was immersed into the mixture in a way that ensured the tip was fully submerged inside the solvent mixture. A similar procedure for ultrasonication as of esterification was followed. The transesterification process was carried out with 1 w/wt% KOH in 6:1 methanol to oil ratio for 30, 45, 60 and 75 min, as a part of the optimisation of reaction time. The temperature was not optimised for the reason explained previously. After transesterification, the separation was carried out by allowing the mixture to settle. The bottom layer (glycerol layer) was removed before washing off the top layer. Washing was done with warm water (60 °C) several times until no impurities were observed in the water. The biodiesel was then placed in a rotary evaporator to remove any remaining moisture. The top biodiesel layer was tested for fatty acid methyl ester (FAME) contents using the Agilent 7890A Gas Chromatograph (GC). Figure 2 shows the phase separation after transesterification.

Figure 2. Phase separation after transesterification.

2.7. The GC Analysis of the Fatty Acid Composition (FAC)

A GC (Agilent 7890A) fitted with a flame ionisation detector was used to determine the FAME content of the produced biodiesel. Carbon chains (C8-C24) in the FAME layer and linolenic acid methyl ester content of the biodiesel were measured following the EN 14103:2011 standard method with methyl nonadecanoate (internal standard). This method is suitable for use with the GC equipped with HP-INNOWax high-polarity column (length × inner diameter × film thickness: 30 m × 0.25 mm × 0.25 μm, stationary phase: polyethylene glycol). The oven temperature protocol: 2 min constant 60 °C, heated to 200 °C at 10 °C increase per minute, then to 240 °C at 5 °C increase per minute and finally 7 min at 240° C. Helium gas was used as the carrier gas, with a flow rate for helium of 1.5 mL/min. All the FAMEs were chromatographically resolved at the approximate retention time of 25 min. The comparison between the area of methyl ester peaks and internal standard peaks provided the FAME yield (Equation (4)):

$$E = \frac{(\sum A) - A_{EI}}{A_{EI}} \times \frac{W_{EI}}{m} \times 100\% \tag{4}$$

where E signifies the percentage FAME content (%), $\sum A$ the sum of the area of C8:0 to C24:0 peaks, A_{EI} the peak area of internal standards, W_{EI} the weight (mg) of methyl nonadecanoate and m is the weight (mg) of the sample.

2.8. FT-IR Analysis

The characterisation of WCG methyl ester was carried out by FT-IR (Perkin Elmer equipped with the MIR TGS detector) in the range 4000–650 cm^{-1} and analysed with the software program 'Spectrum'. The resolution was 8 scans and between 8 cm^{-1} and 4 cm^{-1}.

3. Results and Discussions

3.1. Comparison of Soxhlet and Ultrasonic-Assisted Extraction

From Table 3, the WCG reaction time was found to be 30 min with the highest oil yield obtained being 15.84% using ultrasound-assisted extraction. As for Soxhlet extraction, the highest yield was found to be 15.62% at 180 min reaction time. It is seen that for the constant solvent, ultrasonic-assisted extraction technology offers a higher oil yield and takes less time to complete compared to Soxhlet technology.

Table 3. Comparison of WCG oil yield of Soxhlet and ultrasonic-assisted extraction process.

Method	Time (min)	n-hexane	Yield (%)
	20	1:20	13.73
Ultrasound	30	1:20	15.84
	40	1:20	14.82
	60	1:20	8.9
Soxhlet	120	1:20	12.74
	180	1:20	15.62

3.2. Effect of Solvent on Lipid Extraction Yield

Results on the effect of solvents on lipid extraction showed that n-hexane offers the highest yield (15.8%) followed by, chloroform (12.3%) and methanol (9.11%). Reshad [29] reported improved performance of n-hexane for rubber seed oil extraction. Al-Hammare et al. [30] also obtained a higher oil yield (15.3%) when using the Soxhlet technique and n-hexane solvent. Although the oil yield is higher than that reported in this study, this may be due to the nature of the source from which the WCG was obtained. Oil extraction using methanol resulted in 18.67% more viscous oil than n-hexane which might be the result of non-oil components being extracted together with the oil components due to its polarity, as suggested by Perrier [31]. Further, due to the alcohol's low selectivity to triglycerides, the extraction will involve other compounds such as polyphenols, phosphatides, soluble sugars and pigments [32]. An ethanol and chloroform solvent mixture were found to produce the highest oil yield for microalgae *Chlorella* sp. [33]. This suggests that the selection of solvent for oil extraction depends on the feedstock used. It is to be noted that effective oil extraction requires complete solvent penetration into oil storage and matching targeted compounds polarity [34].

3.3. Optimisation of the WCG Oil Yield Using Response Surface Methodology

In this study, WCG oil yield was maximised by optimising the independent process variables, such as amplitude, reaction time and amount of n-hexane. The quadratic regression model was suggested after a regression analysis performed on Box-Behnken experimental design results. The WCG oil yield results for 17 experimental runs obtained using the quadratic regression model equation are shown in Table 4.

Table 4. Experimental design for the optimisation WCG extraction process.

Run	X_1 Amplitude	X_2 Time	X_3 n-hexane	Experiment Yield	Predicted Yield
1	20	37.5	15	11.6	11.53
2	40	37.5	30	16.44	16.51
3	40	37.5	15	13.03	12.99
4	30	25	30	17.28	17.32
5	40	25	22.5	15.22	15.11
6	20	25	22.5	13.18	13.10
7	30	37.5	22.5	16.42	16.49
8	20	37.5	30	14.22	14.26
9	30	37.5	22.5	16.57	16.49
10	30	37.5	22.5	16.52	16.49
11	20	50	22.5	12.53	12.64
12	30	50	15	13.6	13.57
13	30	25	15	14.13	14.28
14	30	50	30	16.94	16.79
15	30	37.5	22.5	16.38	16.49
16	30	37.5	22.5	16.56	16.49
17	40	50	22.5	14.26	14.34

The WCG oil yield is predicted by the quadratic model in the form of coded values is shown in Equation (5).

$$Y = -17.18125 + 1.364X_1 + 0.229X_2 + 0.5017X_3 - 0.00062X_1X_2 + 0.00263X_1X_3 + 0.000507X_2X_3$$
$$-0.0217875X_1^2 - 0.003288X_2^2 - 0.00869X_3^2 \tag{5}$$

Here, Y shows the WCG oil yield and X_1, X_2 and X_3 exhibit the amplitude, reaction time, and amount of n-hexane.

The significance of the response surface model (quadratic) to optimise the WCG oil yield was evaluated using ANOVA. The results are presented in Table 5. The quadratic model p-value was also <0.0001, which indicated that the quadratic regression model was "significant". Model terms are significant if values of "Prob > F" are less than 0.0500 and model terms are insignificant if values of "Prob > F" are greater than 0.10. The lack of fit "F value" is 0.0927, which indicates that lack of fit was not significant relative to the pure error. The value R2 is 0.9975, which indicates that 99.75% of the deviation in USC crude oil yield was due to independent input process variables chosen for this model. According to high R^2 value, data points will be closer to the regression line, and it was a better estimation between the experimental data and quadratic model.

Table 5. Analysis of variance results for a quadratic regression model.

Source	Sum of Squares	Degree of Freedom	Mean	F-Value	p-Value	
Model	50.81	9	5.65	313.92	< 0.0001	Significant
X_1-Amplitude	6.88	1	6.88	382.64	< 0.0001	
X_2-Time	0.7688	1	0.7688	42.75	0.0003	
X_3-n-hexane	19.59	1	19.59	1089.41	< 0.0001	
X_1X_2	0.0240	1	0.0240	1.34	0.2857	
X_1X_3	0.1560	1	0.1560	8.67	0.0215	
X_2X_3	0.0090	1	0.0090	0.5018	0.5016	
X_1^2	19.99	1	19.99	1111.28	< 0.0001	
X_2^2	1.11	1	1.11	61.79	0.0001	
X_3^2	1.01	1	1.01	55.92	0.0001	
Residual	0.1259	7	0.0180			
Lack of Fit	0.0967	3	0.0322	4.42	0.0927	not significant
Pure Error	0.0292	4	0.0073			
Cor Total	50.94	16				

The calculated yield of WCG oil at the optimal condition of 1:30 g/g of the mass of oil: n-hexane for 34 min of reaction time at the 32% amplitude was 17.75 wt. %. The yield of calculated coffee oil under the optimal condition can be proved by the experiment. The results from the experiment showed that 17.23 wt. % of coffee oil can be extracted from WCG under the optimal condition. Thus, the experimental yield closely matches the calculated coffee oil yield.

3.3.1. Effect of Ultrasonic Amplitude and Period on WCG Lipid Yield

As ultrasonication amplitude increases bubble collapse becomes more violent, which leads to higher extraction yields. However, an excessive ultrasonic amplitude may lead to wastage of energy. The WCG oil extraction versus ultrasonic amplitude and period is shown in Figure 3a,b. With the use of n-hexane solvent, the increase of amplitude from 20% to 33% resulted in a significant increase in oil extraction. The ultrasonic power increase in this range enhances the molecular diffusion of oil into the n-hexane. However, the extraction yield decreases after this amplitude. Liu et al. [35] reported a reduction of oil yield after a specific ultrasonic power threshold and attributed this to intense heating resulting in decomposition and volatility of the oil. The optimum condition resulted in an oil extraction of 17.75 wt. % using ultrasonic assistance, significantly higher than that obtained using the Soxhlet extraction method. For rapeseed oil extraction, Sicaire et al. [36] reported that the chief contributors to optimised oil yield were ultrasonic intensity and solvent-to-solid ratio. They also reported a reduction of oil yield when ultrasonic power exceeded the optimum values (> 30%).

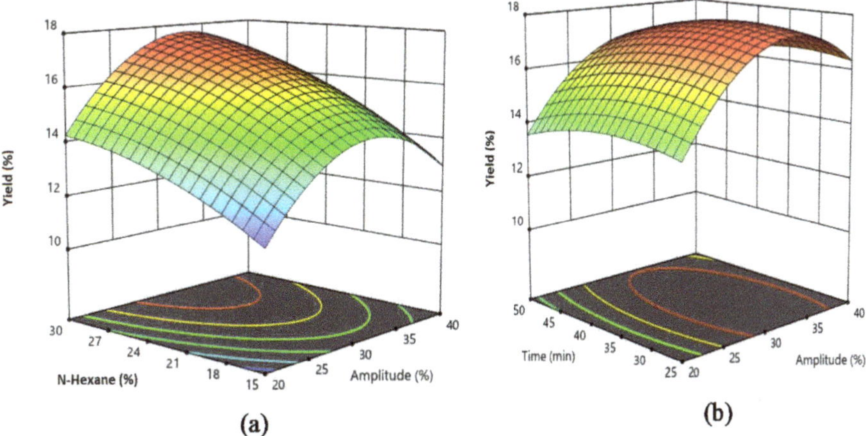

Figure 3. Three-dimensional surface plot of (**a**) n-hexane vs amplitude, (**b**) time vs amplitude.

The optimum extraction period avoids excessive power use and reduces the risk of oil damage. Sufficient extraction time is required to break ground coffee cell walls, extract the lipids and achieve an equilibrium [37]. From the results, the optimum ultrasonic oil extraction period using n-hexane was 34 min, significantly lower than the 3 h required by the Soxhlet extraction method for the same solvent. Thus, the oil is not subjected to intense heating for long periods which reduces the risk of oil degradation [38]. Zhang et al. [39] reported a similar optimum extraction period for flaxseed oil. The initial rapid oil extraction rate is due to the solvent penetration into the cellular structure at a faster rate. Due to the oil constituents' external diffusion through the porous residual solids, a reduced extraction rate at a later stage was observed [40]. Evaporation might have resulted in the loss of a large portion of solvent which reduced the extraction efficiency. WCG cell wall rupture results in suspended impurities within the extract which may reduce the permeability of solvent into cell structure [41]. Furthermore, the extended periods of ultrasonication might also result in WCG oil re-adsorbing into ruptured tissue particles due to the larger specific surface area [42].

3.3.2. Effect of Amount of n-hexane on Lipid Extraction Yield

Figure 4 shows a 3D plot of the reaction time versus the amount of n-hexane. From the results, the optimum n-hexane for WCG ultrasonic oil extraction is 1:30. High extraction rates require a high amount of n-hexane, but extended extraction periods, in turn, increase energy consumption as well as the chance of degradation of extracted oil.

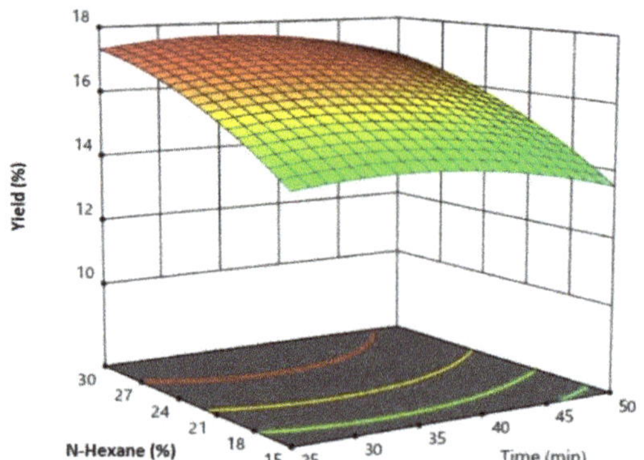

Figure 4. Three-dimensional surface plot of n-hexane vs time.

3.4. Properties of Recovered Lipid

The acid value of ultrasonically extracted WCG oil was 9.56 mg KOH/g oil. The viscosity and density of the oil were determined to be 43.6 mm^2/s and 918.4 kg/m^3 respectively. The WCG oil higher heating value (HHV) was measured to be 38.85 MJ/kg, exceeding the HHV of other biodiesel feedstocks such as waste cooking oil [43], cottonseed oil [44] and rice bran oil [45]. The higher calorific value might be due to the high carbon to oxygen ratios [46].

3.5. Esterification and Transesterification Process

To reduce the total energy required for oil extraction, the ultrasonic amplitude was optimised. Figure 5 depicts WCG fatty acid methylester (FAME) yield after different periods. The FAME yield continues to increase up to 60 min reaction time reaching a peak (98.21%) before slightly reducing beyond 60 min reaction time. Ultrasonic assistance is known to reduce the reaction time for biodiesel conversion. The 60 min reaction time required using ultrasonic assistance is significantly lower compared to conventional solvent transesterification of WCG (12 h for optimum conversion) [35]. Temperature and time affect the transesterification reaction as it is kinetically controlled. Temperature changes during long sonication periods have been known to destroy the oil [47]. It is also challenging to maintain the temperature throughout the ultrasonication process due to the constant compression and rarefactions, resulting in a temperature which fluctuates. By controlling the reaction period, the total amount of energy introduced into the system can be easily quantified. Furthermore, given the large heat capacities of certain oils, the effect temperature has during the transesterification reaction is small when compared to the time [48]. Extended reaction times would decrease FAME yields due to degradation and polymerisation [49].

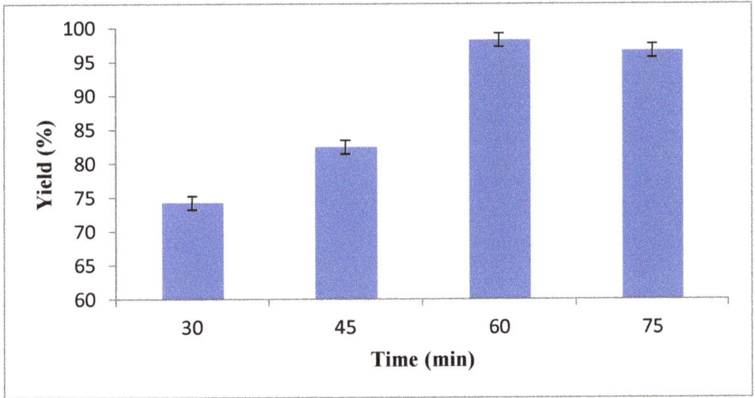

Figure 5. Relationship between biodiesel yield and reaction time.

3.6. Characterisation of FAME of WCG

3.6.1. GC Analysis

The FAC was determined via GC following the steps described in the methodology section. WCG biodiesel is known to have a similar FAC to corn and soybean biodiesel [50]. The FAME profile for the biodiesel produced optimally is reported in Table 6.

Table 6. FAC of WCG biodiesel.

Fatty Acids	Molecular Weight	Structure	Formula	wt.%
Myristic acid	228	14:0	$C_{14}H_{28}O_2$	3.82
Myristoleic	226	14:1	$C_{14}H_{26}O_2$	20
Palmitic	256	16:0	$C_{16}H_{32}O_2$	19
Stearic	284	18:0	$C_{18}H_{36}O_2$	6.73
Oleic	282	18:1	$C_{18}H_{34}O_2$	9.27
Linoleic	280	18:2	$C_{18}H_{32}O_2$	28.71
Arachidic	312	20:0	$C_{20}H_{40}O_2$	2.96
Tricosanoic	338	23:0	$C_{23}H_{46}O_2$	5.11
Lignoceric	368	24:1	$C_{24}H_{46}O_2$	4.29
Total saturated fatty acid				37.61
Total monounsaturated fatty acid (MUFA)				33.56
Total polyunsaturated fatty acid (PUFA)				28.71

As shown in Table 6, WCG biodiesel has 37.61% saturated fatty acids and 62.27% unsaturated fatty acids, the latter of which consists of 33.56% monounsaturated fatty acids and 28.71% polyunsaturated fatty acids. Among the fatty acids linoleic (18:2) is the predominant fatty acid compared to other fatty acids.

3.6.2. FT-IR Analysis

The characteristic peaks of the WCG methyl ester are found at 2926 cm^{-1}, 2854 cm^{-1}, 1743 cm^{-1}, 1435-1463 cm^{-1} and 1163 cm^{-1}, which corresponds to the C–H stretching vibration with strong absorption intensity, CH2 asymmetric and symmetric vibration with strong absorption intensity, C = O stretching vibration with strong absorption intensity, CH2 shear-type vibration with mild absorption intensity, and C–O–C symmetric stretching vibration with mild absorption intensity, respectively. Table 7 shows the characteristic peaks of WCG methyl ester.

Table 7. Characteristic peaks of WCG in FT-IR spectra.

Absorption Bands (cm^{-1})	Functional Group	Absorption Intensity
2926	C–H stretching vibration	Strong
2854	CH2 asymmetric and symmetric vibration	Strong
1743	C = O stretching vibration	Strong
1435–1463	CH2 shear type vibration	Middling
1163	C–O–C symmetric stretching vibration	Middling

3.6.3. Physicochemical Property Analysis

The fuel properties of WCG biodiesel produced through ultrasonic transesterification were analysed and compared to the ASTM D6751 and EN 14214 standards [51]. Table 8 shows the fuel properties of WCG FAME. It was found that the kinematic viscosity of WCG FAME was 4.89 mm^2/s, the higher heating value was 39.74 MJ/kg, density was 886.8 kg/m^3 and the acid value lowered to 0.52 mgKOH/g oil. Further, the ester content was found at 98.21%. However, all these results except the acid value were within the specified limit of the ASTM D6751 and EN14214 standards.

Table 8. Properties of WCG methyl ester.

Properties	WCG Biodiesel	ASTM D6751	EN 14214
Ester content (%)	97	–	Minimum 96.5
Acid value (mg KOH g^{-1})	0.52	Maximum 0.5	Maximum 0.5
Kinematic viscosity at 40 °C (mm^2/s)	4.89	1.9 to 6.0	3.5 to 5.0
Density at 15° C (kg/m^3)	886.8	880	860 to 900
Higher heating value (MJ/kg)	39.74	–	–

4. Conclusions

This paper evaluated the recovery of an energy source from a wasted product, i.e., waste coffee grounds, using ultrasonic assistance to produce biodiesel. Use of this feedstock is advantageous as it is abundant, utilises a food waste, and eliminates the side effects of landfill disposal. Some of the salient results are:

Successful production of biofuel oil from the UCG using ultrasonic assistance which was later converted into biodiesel.

The optimum ultrasonic oil extraction conditions were 1:30 g/g of the mass ratio of oil-to-n-hexane, 32% ultrasonic amplitude with a reaction time of 34 min and an achieved oil yield of 17.75 wt.%.

Ultrasonication of WCG during oil extraction reduced the amount of solvent required significantly. This also reduced extraction time and increased extractability compared to the conventional Soxhlet extraction method.

Thus, the study can suggest ultrasonic assistance is a superior method compared to the Soxhlet extraction method. Furthermore, it is also possible to convert waste-to-energy by producing biodiesel from WCG. Further studies are required to evaluate the produced biodiesel as a substitute for petro-diesel.

Author Contributions: Conceptualization, M.M.; methodology, F.K.; formal analysis, I.M.R.F.; investigation, F.K.; writing—original draft preparation, M.M.; writing—review and editing, H.M.M. and M.G.R.; supervision, T.M.I.M. and A.H.S. All authors have read and agreed to the published version of the manuscript.

Funding: This research was funded by research development fund of School of Information, Systems and Modelling, University of Technology Sydney, Australia. This research also received financial support from AAIBE Chair of Renewable Energy (Grant no: 201801 KETTHA).

Conflicts of Interest: The authors declare no conflict of interest.

References

1. Bhuiya, M.M.K.; Rasul, M.; Khan, M.; Ashwath, N.; Mofijur, M. Comparison of oil extraction between screw press and solvent (n-hexane) extraction technique from beauty leaf (Calophyllum inophyllum L.) feedstock. *Ind. Crops Prod.* **2020**, *144*, 112024. [CrossRef]
2. Mofijur, M.; Mahlia, T.M.I.; Silitonga, A.S.; Ong, H.C.; Silakhori, M.; Hasan, M.H.; Putra, N.; Rahman, S.M.A. Phase Change Materials (PCM) for Solar Energy Usages and Storage: An Overview. *Energies* **2019**, *12*, 3167. [CrossRef]
3. Uddin, M.; Techato, K.; Taweekun, J.; Rahman, M.; Rasul, M.; Mahlia, T.; Ashrafur, S. An Overview of Recent Developments in Biomass Pyrolysis Technologies. *Energies* **2018**, *11*, 3115. [CrossRef]
4. Mofijur, M.; Masjuki, H.H.; Kalam, M.A.; Atabani, A.E. Evaluation of biodiesel blending, engine performance and emissions characteristics of Jatropha curcas methyl ester: Malaysian perspective. *Energy* **2013**, *55*, 879–887. [CrossRef]
5. Mofijur, M.; Masjuki, H.H.; Kalam, M.A.; Atabani, A.E.; Fattah, I.M.R.; Mobarak, H.M. Comparative evaluation of performance and emission characteristics of Moringa oleifera and Palm oil based biodiesel in a diesel engine. *Ind. Crops Prod.* **2014**, *53*, 78–84. [CrossRef]
6. Shahabuddin, M.; Liaquat, A.M.; Masjuki, H.H.; Kalam, M.A.; Mofijur, M. Ignition delay, combustion and emission characteristics of diesel engine fueled with biodiesel. *Renew. Sustain. Energy Rev.* **2013**, *21*, 623–632. [CrossRef]
7. Mahlia, T.M.I.; Syazmi, Z.A.H.S.; Mofijur, M.; Abas, A.E.P.; Bilad, M.R.; Ong, H.C.; Silitonga, A.S. Patent landscape review on biodiesel production: Technology updates. *Renew. Sustain. Energy Rev.* **2020**, *118*, 109526. [CrossRef]
8. Ong, H.C.; Masjuki, H.H.; Mahlia, T.M.I.; Silitonga, A.S.; Chong, W.T.; Yusaf, T. Engine performance and emissions using Jatropha curcas, Ceiba pentandra and Calophyllum inophyllum biodiesel in a CI diesel engine. *Energy* **2014**, *69*, 427–445. [CrossRef]
9. Ong, H.C.; Milano, J.; Silitonga, A.S.; Hassan, M.H.; Shamsuddin, A.H.; Wang, C.T.; Mahlia, T.M.I.; Siswantoro, J.; Kusumo, F.; Sutrisno, J. Biodiesel production from Calophyllum inophyllum-Ceiba pentandra oil mixture: Optimization and characterization. *J. Clean. Prod.* **2019**, *219*, 183–198. [CrossRef]
10. Silitonga, A.; Shamsuddin, A.; Mahlia, T.; Milano, J.; Kusumo, F.; Siswantoro, J.; Dharma, S.; Sebayang, A.; Masjuki, H.; Ong, H.C. Biodiesel synthesis from Ceiba pentandra oil by microwave irradiation-assisted transesterification: ELM modeling and optimization. *Renew. Energy* **2020**, *146*, 1278–1291. [CrossRef]
11. Silitonga, A.S.; Masjuki, H.H.; Mahlia, T.M.I.; Ong, H.C.; Chong, W.T.; Boosroh, M.H. Overview properties of biodiesel diesel blends from edible and non-edible feedstock. *Renew. Sustain. Energy Rev.* **2013**, *22*, 346–360.
12. Fazeli Danesh, A.; Ebrahimi, S.; Salehi, A.; Parsa, A. Impact of nutrient starvation on intracellular biochemicals and calorific value of mixed microalgae. *Biochem. Eng. J.* **2017**, *125*, 56–64. [CrossRef]
13. Döhlert, P.; Weidauer, M.; Enthaler, S. Spent coffee ground as source for hydrocarbon fuels. *J. Energy Chem.* **2016**, *25*, 146–152. [CrossRef]
14. Allesina, G.; Pedrazzi, S.; Allegretti, F.; Tartarini, P. Spent coffee grounds as heat source for coffee roasting plants: Experimental validation and case study. *Appl. Therm. Eng.* **2017**, *126*, 730–736. [CrossRef]
15. Murthy, P.S.; Madhava Naidu, M. Sustainable management of coffee industry by-products and value addition—A review. *Resour. Conserv. Recycl.* **2012**, *66*, 45–58. [CrossRef]
16. Zabaniotou, A.; Kamaterou, P. Food waste valorization advocating Circular Bioeconomy—A critical review of potentialities and perspectives of spent coffee grounds biorefinery. *J. Clean. Prod.* **2019**, *211*, 1553–1566. [CrossRef]
17. Cameron, A.; O'Malley, S. Coffee Ground Recovery Program Summary Report. Available online: https://planetark.org/documents/doc-1397-summary-report-of-feasibility-study-april-2016.pdf (accessed on 30 December 2019).
18. Kookos, I.K. Technoeconomic and environmental assessment of a process for biodiesel production from spent coffee grounds (SCGs). *Resour. Conserv. Recycl.* **2018**, *134*, 156–164. [CrossRef]
19. Palconite, C.L.; Edrolin, A.C.; Lustre, S.N.B.; Manto, A.A.; Caballero, J.R.L.; Tizo, M.S.; Ido, A.L.; Arazo, R.O. Optimization and characterization of bio-oil produced from Ricinus communis seeds via ultrasonic-assisted solvent extraction through response surface methodology. *Sustain. Environ. Res.* **2018**, *28*, 444–453. [CrossRef]

20. Gui, M.M.; Lee, K.; Bhatia, S. Feasibility of edible oil vs. non-edible oil vs. waste edible oil as biodiesel feedstock. *Energy* **2008**, *33*, 1646–1653. [CrossRef]

21. Atabani, A.E.; Shobana, S.; Mohammed, M.N.; Uğuz, G.; Kumar, G.; Arvindnarayan, S.; Aslam, M.; Al-Muhtaseb, A.A.H. Integrated valorization of waste cooking oil and spent coffee grounds for biodiesel production: Blending with higher alcohols, FT–IR, TGA, DSC and NMR characterizations. *Fuel* **2019**, *244*, 419–430. [CrossRef]

22. Dharma, S.; Masjuki, H.H.; Ong, H.C.; Sebayang, A.H.; Silitonga, A.S.; Kusumo, F.; Mahlia, T.M.I. Optimization of biodiesel production process for mixed Jatropha curcas-Ceiba pentandra biodiesel using response surface methodology. *Energy Convers. Manag.* **2016**, *115*, 178–190. [CrossRef]

23. Jamaluddin, N.A.M.; Riayatsyah, T.M.I.; Silitonga, A.S.; Mofijur, M.; Shamsuddin, A.H.; Ong, H.C.; Mahlia, T.M.I.; Rahman, S.M.A. Techno-economic analysis and physicochemical properties of Ceiba pentandra as second-generation biodiesel based on ASTM D6751 and EN 14214. *Processes* **2019**, *7*, 636. [CrossRef]

24. Coh, B.H.H.; Ong, H.C.; Cheah, M.Y.; Chen, W.-H.; Yu, K.L.; Mahlia, T.M.I. Sustainability of direct biodiesel synthesis from microalgae biomass: A critical review. *Renew. Sustain. Energy Rev.* **2019**, *107*, 59–74.

25. Silitonga, A.S.; Mahlia, T.M.I.; Kusumo, F.; Dharma, S.; Sebayang, A.H.; Sembiring, R.W.; Shamsuddin, A.H. Intensification of Reutealis trisperma biodiesel production using infrared radiation: Simulation, optimisation and validation. *Renew. Energy* **2019**, *133*, 520–527. [CrossRef]

26. Hashemi, S.M.B.; Michiels, J.; Asadi Yousefabad, S.H.; Hosseini, M. Kolkhoung (Pistacia khinjuk) kernel oil quality is affected by different parameters in pulsed ultrasound-assisted solvent extraction. *Ind. Crops Prod.* **2015**, *70*, 28–33. [CrossRef]

27. Tan, C.H.; Nagarajan, D.; Show, P.L.; Chang, J.-S. Chapter 25—Biodiesel From Microalgae. In *Biofuels: Alternative Feedstocks and Conversion Processes for the Production of Liquid and Gaseous Biofuels*, 2nd ed.; Pandey, A., Larroche, C., Dussap, C.-G., Gnansounou, E., Khanal, S.K., Ricke, S., Eds.; Academic Press: Cambridge, MA, USA, 2019; pp. 601–628.

28. Rocha, M.V.P.; de Matos, L.J.B.L.; Lima, L.P.d.; Figueiredo, P.M.d.S.; Lucena, I.L.; Fernandes, F.A.N.; Gonçalves, L.R.B. Ultrasound-assisted production of biodiesel and ethanol from spent coffee grounds. *Bioresour. Technol.* **2014**, *167*, 343–348. [CrossRef]

29. Reshad, A.S.; Tiwari, P.; Goud, V.V. Extraction of oil from rubber seeds for biodiesel application: Optimization of parameters. *Fuel* **2015**, *150*, 636–644. [CrossRef]

30. Al-Hamamre, Z.; Foerster, S.; Hartmann, F.; Kröger, M.; Kaltschmitt, M. Oil extracted from spent coffee grounds as a renewable source for fatty acid methyl ester manufacturing. *Fuel* **2012**, *96*, 70–76. [CrossRef]

31. Perrier, A.; Delsart, C.; Boussetta, N.; Grimi, N.; Citeau, M.; Vorobiev, E. Effect of ultrasound and green solvents addition on the oil extraction efficiency from rapeseed flakes. *Ultrason. Sonochemistry* **2017**, *39*, 58–65. [CrossRef]

32. Baümler, E.R.; Carrín, M.E.; Carelli, A.A. Extraction of sunflower oil using ethanol as solvent. *J. Food Eng.* **2016**, *178*, 190–197. [CrossRef]

33. Ramluckan, K.; Moodley, K.G.; Bux, F. An evaluation of the efficacy of using selected solvents for the extraction of lipids from algal biomass by the soxhlet extraction method. *Fuel* **2014**, *116*, 103–108. [CrossRef]

34. Adeoti, I.A.; Hawboldt, K. A review of lipid extraction from fish processing by-product for use as a biofuel. *Biomass Bioenergy* **2014**, *63*, 330–340. [CrossRef]

35. Liu, P.; Xu, Y.-f.; Gao, X.-d.; Zhu, X.-y.; Du, M.-z.; Wang, Y.-x.; Deng, R.-x.; Gao, J.-y. Optimization of ultrasonic-assisted extraction of oil from the seed kernels and isolation of monoterpene glycosides from the oil residue of Paeonia lactiflora Pall. *Ind. Crops Prod.* **2017**, *107*, 260–270. [CrossRef]

36. Sicaire, A.-G.; Vian, M.A.; Fine, F.; Carré, P.; Tostain, S.; Chemat, F. Ultrasound induced green solvent extraction of oil from oleaginous seeds. *Ultrason. Sonochemistry* **2016**, *31*, 319–329. [CrossRef] [PubMed]

37. Son, J.; Kim, B.; Park, J.; Yang, J.; Lee, J.W. Wet in situ transesterification of spent coffee grounds with supercritical methanol for the production of biodiesel. *Bioresour. Technol.* **2018**, *259*, 465–468. [CrossRef]

38. Chen, X.; Luo, Y.; Qi, B.; Wan, Y. Simultaneous extraction of oil and soy isoflavones from soy sauce residue using ultrasonic-assisted two-phase solvent extraction technology. *Sep. Purif. Technol.* **2014**, *128*, 72–79. [CrossRef]

39. Zhang, Z.-S.; Wang, L.-J.; Li, D.; Jiao, S.-S.; Chen, X.D.; Mao, Z.-H. Ultrasound-assisted extraction of oil from flaxseed. *Sep. Purif. Technol.* **2008**, *62*, 192–198. [CrossRef]

40. Goula, A.M. Ultrasound-assisted extraction of pomegranate seed oil—Kinetic modeling. *J. Food Eng.* **2013**, *117*, 492–498. [CrossRef]

41. Tian, Y.; Xu, Z.; Zheng, B.; Martin Lo, Y. Optimization of ultrasonic-assisted extraction of pomegranate (Punica granatum L.) seed oil. *Ultrason. Sonochemistry* **2013**, *20*, 202–208. [CrossRef]

42. Dong, J.; Liu, Y.; Liang, Z.; Wang, W. Investigation on ultrasound-assisted extraction of salvianolic acid B from Salvia miltiorrhiza root. *Ultrason. Sonochemistry* **2010**, *17*, 61–65. [CrossRef]

43. Madheshiya, A.K.; Vedrtnam, A. Energy-exergy analysis of biodiesel fuels produced from waste cooking oil and mustard oil. *Fuel* **2018**, *214*, 386–408. [CrossRef]

44. Shankar, A.A.; Pentapati, P.R.; Prasad, R.K. Biodiesel synthesis from cottonseed oil using homogeneous alkali catalyst and using heterogeneous multi walled carbon nanotubes: Characterization and blending studies. *Egypt. J. Pet.* **2017**, *26*, 125–133. [CrossRef]

45. Mazaheri, H.; Ong, H.C.; Masjuki, H.H.; Amini, Z.; Harrison, M.D.; Wang, C.-T.; Kusumo, F.; Alwi, A. Rice bran oil based biodiesel production using calcium oxide catalyst derived from Chicoreus brunneus shell. *Energy* **2018**, *144*, 10–19. [CrossRef]

46. Kelkar, S.; Saffron, C.M.; Chai, L.; Bovee, J.; Stuecken, T.R.; Garedew, M.; Li, Z.; Kriegel, R.M. Pyrolysis of spent coffee grounds using a screw-conveyor reactor. *Fuel Process. Technol.* **2015**, *137*, 170–178. [CrossRef]

47. Bahmani, L.; Aboonajmi, M.; Arabhosseini, A.; Mirsaeedghazi, H. Effects of ultrasound pre-treatment on quantity and quality of essential oil of tarragon (Artemisia dracunculus L.) leaves. *J. Appl. Res. Med. Aromat. Plants* **2018**, *8*, 47–52. [CrossRef]

48. Lee, S.B.; Lee, J.D.; Hong, I.K. Ultrasonic energy effect on vegetable oil based biodiesel synthetic process. *J. Ind. Eng. Chem.* **2011**, *17*, 138–143. [CrossRef]

49. Thiruvenkadam, S.; Izhar, S.; Hiroyuki, Y.; Harun, R. One-step microalgal biodiesel production from Chlorella pyrenoidosa using subcritical methanol extraction (SCM) technology. *Biomass Bioenergy* **2019**, *120*, 265–272. [CrossRef]

50. Efthymiopoulos, I.; Hellier, P.; Ladommatos, N.; Mills-Lamptey, B. Transesterification of high-acidity spent coffee ground oil and subsequent combustion and emissions characteristics in a compression-ignition engine. *Fuel* **2019**, *247*, 257–271. [CrossRef]

51. Silitonga, A.S.; Masjuki, H.H.; Ong, H.C.; Sebayang, A.H.; Dharma, S.; Kusumo, F.; Siswantoro, J.; Milano, J.; Daud, K.; Mahlia, T.M.I.; et al. Evaluation of the engine performance and exhaust emissions of biodiesel-bioethanol-diesel blends using kernel-based extreme learning machine. *Energy* **2018**, *159*, 1075–1087. [CrossRef]

Review

Current Research and Development Status of Corrosion Behavior of Automotive Materials in Biofuels

Aamir Shehzad [1], Arslan Ahmed [1], Moinuddin Mohammed Quazi [2], Muhammad Jamshaid [3], S. M. Ashrafur Rahman [4,*], Masjuki Haji Hassan [5] and Hafiz Muhammad Asif Javed [6]

[1] Department of Mechanical Engineering, COMSATS University Islamabad, Sahiwal Campus 57000, Pakistan; aamir.shehzad@cuisahiwal.edu.pk (A.S.); arslanahmad@cuisahiwal.edu.pk (A.A.)

[2] Faculty of Mechanical and Automotive Engineering Technology, Universiti Malaysia Pahang, Pekan 26600, Pahang, Malaysia; moinuddin@ump.edu.my

[3] Department of Mechanical Engineering, Bahauddin Zakariya University, Multan 60000, Pakistan; muhammad.jamshaid@bzu.edu.pk

[4] Biofuel Engine Research Facility, Queensland University of Technology, Brisbane, QLD 4000, Australia

[5] Department of Mechanical Engineering, Faculty of Engineering, International Islamic University Malaysia (IIUM), Jalan Gombak, Kuala Lumpur 53100, Malaysia; masjuki@iium.edu.my

[6] Department of Physics, University of Agriculture Faisalabad, Faisalabad 38000, Pakistan; majavedphy@yahoo.com

* Correspondence: s2.rahman@qut.edu.au

Citation: Shehzad, A.; Ahmed, A.; Quazi, M.M.; Jamshaid, M.; Ashrafur Rahman, S.M.; Hassan, M.H.; Javed, H.M.A. Current Research and Development Status of Corrosion Behavior of Automotive Materials in Biofuels. *Energies* **2021**, *14*, 1440. https://doi.org/10.3390/en14051440

Academic Editor: Constantine D. Rakopoulos

Received: 30 November 2020
Accepted: 22 February 2021
Published: 6 March 2021

Abstract: The world's need for energy is increasing with the passage of time and the substantial energy demand of the world is met by fossil fuels. Biodiesel has been considered as a replacement for fossil fuels in automotive engines. Biodiesels are advantageous because they provide energy security, they are nontoxic, renewable, economical, and biodegradable and clean sources of energy. However, there are certain disadvantages of biodiesels, including their corrosive, hygroscopic and oxidative natures. This paper provides a review of automotive materials when coming into contact with biodiesel blended fuel in terms of corrosion. Biodiesels have generally been proved to be corrosive, therefore it is important to understand the limits and extents of corrosion on different materials. Methods generally used to find and calculate corrosion have also been discussed in this paper. The reasons for the occurrence of corrosion and the subsequent problems because of corrosion have been presented. Biodiesel production can be carried out by different feedstocks and the studies which have been carried out on these biodiesels have been reviewed in this paper. A certain number of compounds form on the surface of materials because of corrosion and the mechanism behind the formation of these compounds along with the characterization techniques generally used is reviewed.

Keywords: corrosion; biodiesel; automotive materials; green fuels; corrosion test methods

1. Introduction

In biodiesel, the term "bio" implies that it is renewable as compared with petroleum fuels, and "diesel" represents that it has the potential to be used in diesel cycle motors [1]. Biodiesel is composed of unsaturated and saturated ester components, because of which it is considered to be less stable [2] and sensitive to light [3], temperature and metal ions. It can be obtained from animal fat, used cooking oil and vegetable oil with the help of methanol and ethanol [4]. The common sources of biofuels are animal fats and vegetables [5]. The common feedstock used for biodiesel production includes palm oil [6], sunflower oil [7], rapeseed oil [8], canola oil [9], soybean oil [10] and corn oil [11]. Another source of biodiesel production is waste chicken fat oil [12]. A transesterification process is applied to vegetable oils to produce biodiesel [13]. Some of the properties of both diesel and biodiesel are similar; however, they differ from each other because of chemical variations. Petroleum diesel has hundreds of compounds [14] that boil at different temperatures, while on the other hand biodiesel has only a few compounds, some of which are esters of long chain alkyls. Biodiesel has wastes that have resulted from the transesterification reactions of methanol, mono- and diglycerides, triglyceride intermediates and fatty acid derivatives other than the main constituents [15].

The world's need for energy is increasing over time because of increasing population and technological development [16]. Fossil fuels are meeting the substantial energy demand of the world, but are expensive [17]. The large production of fossil fuels is leading towards their depletion, which is why the world is going to suffer a huge energy crisis [18]. The extensive use of fossil fuels is leading to global warming and if not controlled, it can raise the temperature of the earth [19]. In addition to this, petroleum products obtained from fossil fuels are polluting the environment [20]. Thus, there is a need for an alternate source of energy. The importance of biodiesel has increased because of its ability to be used as an alternative fuel [21]. Moreover, biofuels are attractive because of their advantages such as being renewable [22], biodegradable [23], economical [24] and nontoxic [25]. Biodiesel can provide energy security to the world whilst having independence from fossil fuels [26], it can reduce emissions and it is a clean source of energy [27], it can be used directly or as a blend in diesel engines [28], it has a higher cetane number [29] and flash point [30] and additionally it showed improved combustion [31] and lubricity [32]. There are certain disadvantages associated with biodiesel as well. Such as its corrosive nature [33], it is more hygroscopic [34] and provides a slightly lower engine performance [35]. Furthermore, the fuel consumption of the engine increases when biodiesel is used [36], and the wear rate of parts in biodiesel is slightly increased [37]. Biodiesel is less volatile [38] and has poor properties in low temperatures [39].

Biofuel usage in the transport sector has been started as a replacement for gasoline [40]. Biodiesel is gaining importance because of its property of direct usage in engines or as a blending component in engines. Different countries of the world have started using it as a blend with petroleum diesel [41]. Research has shown that 30% of ethanol with biodiesel can give effective results when added to diesel fuel. Additionally, it was recommended that 10% ethanol and 20% biodiesel can give better engine performance and the fewest emissions [42].

The biofuels which have been used in engines include palm oil [43], soybean oil [44], rapeseed oil [45], sunflower oil [46], olive oil [47], castor oil [48], jatropha curcas oil [49], pongamia pinnata oil [50], linseed oil [51] and milkweed seed oil [52], etc. Different parts of the engine are made from different materials. Some of the most common parts of the engines include exhaust system, piston assembly, fuel pump, fuel filter, fuel fed-up, fuel tank and fuel injection cylinder; the most common materials which are used in the manufacture of these parts include steel, aluminum, copper, plastic, rubber and ceramic fiber [53]. Engine parts including piston rings, pistons, bearings, filters, fuel injector, fuel liners, gaskets and fuel pump come in contact with the fuel [54]. Among these parts, copper-based alloys become most affected by the fuel [55]. The use of biofuels has some favorable effects on the material of the engine [56]. Biofuels have better lubricity at room temperature and by increasing the concentration of biofuel, deformation of worn surface decreases [37].

In addition to the introduction and pros and cons of biodiesel, this review has described the corrosion studies carried out on biodiesels obtained from feedstocks of different origins on engine materials. Furthermore, methods used to assess corrosion have been discussed in this paper. Other than that, the reasons and problems of corrosion are elaborated in this paper. Additionally, the characterization techniques along with the findings in corrosion studies and the mechanisms of products obtained after the corrosion on the materials have been discussed in this paper.

2. Methods to Find Corrosion

There are two methods that have been adopted by researchers to find the corrosion of materials in contact with biodiesel.

2.1. Immersion Test Method

The process used is according to the ASTM G1 standard and it starts with the cutting of materials and inserting a hole in the desired dimensions. This hole is used to hang the material in the fuel. The materials are then ground and polished with the help of silicon carbide papers of different grades [57]. The samples are then washed with deionized water, dried and then dipped in acetone to degrease them [58]. The samples of each material are weighed initially on a digital balance with accuracy up to four decimal places. The samples are then immersed in the fuels for the specified duration, as shown in Figure 1. After removal, the materials are again washed and degreased by using acetone and then weighed again [41]. The difference in the weight of initial and final samples is further used in the calculation of the Corrosion Rate (CR) of materials according to the following formula [59].

$$CR = (8.76 \times 10^4 \times W)/(D \times A \times T) \text{ mm/year} \tag{1}$$

where:

W = weight loss (g);
D = density (g/cm^3);
A = cross sectional area (cm^2);
T = time (hours).

Figure 1. Schematic diagram for immersion test [1].

2.2. Electro-Chemical Method

This method uses the process of reduction and oxidation reactions. When corrosion occurs, the metal oxidizes and is gained or reduced in the solution. As this reaction involves the flow of electrons and current, it can be measured and calculated electronically [60].

The immersion test method is used generally by most researchers and the reason behind using this method is the efficiency of results [61], as the results obtained by this method use the actual scenario of contact between the biodiesel and the materials. Additionally, this testing is conducted for longer durations while the electrochemical method is carried out for only a few hours. Therefore, a longer duration allows the materials to settle properly in the fuel and results obtained by this method show linear relations. Additionally, this method helps in identifying products of corrosion obtained after the testing by using some characterization techniques. Moreover, the color of fuel and samples change in this method which helps in identifying the compositional changes of the fuel because of the corrosion. Hence, this method is preferred.

3. Reasons and Problems of Corrosion of Materials in Biodiesel

Biodiesel is believed to be corrosive and the reason behind this is its degradation, which is caused by the oxidation reactions taking place because of the absorption of moisture. The corrosiveness of biodiesel becomes more intense if it contains free fatty acids and free water. Likewise, the corrosiveness of biofuels can be increased by auto-oxidation [55]. The feedstock with which biodiesel is synthesized affects the corrosion of metals and this is because of the variation in the chemical composition of the biodiesel obtained from different feedstocks [62]. The rate of corrosion is promoted because the water condenses on the surface of materials [63].

The problem with biodiesel is its degradation when it is exposed to moisture and it oxidizes [64]. Metal oxides form because of biodiesel oxidation [1]. As the corrosiveness of biodiesel is higher, its wear rate becomes higher [55]. When the composition of the fuel changes or when an alternative fuel is used in engines, there comes the issues of material degradation and compatibility of the material with the fuel [42]. Biodiesel usage in the automotive sector is not particularly due to their corrosiveness and degradation properties [41]. Because of the absorption of moisture by the fuel or oxidation, the corrosion damage to the fuel system parts becomes more accelerated. Corrosion and wear of engine parts are increased by the oxidative behavior of the biodiesel [63].

4. Corrosiveness of Biodiesels Obtained from Different Feedstock

Different biodiesels have been used by different researchers for the corrosion rate calculation. The details of those studies are given below.

4.1. Corrosion Studies on Palm Oil-Based Biodiesel

Various researchers conducted studies to find the corrosion rate of automotive materials by using palm oil biodiesel. In their study, Thangavelu et al. investigated the corrosion of copper in Biodiesel-Diesel-Ethanol (BDE) fuel at Room Temperature (RT) and 50 °C by immersion tests. These tests were performed for 408 h. The blends used were 45% biodiesel, 35% diesel and 20% bioethanol (B45D35E20). After 408 h, at room temperature the corrosion rate was 0.277 mpy and at 50 °C it was 0.327 mpy. Pitting on the copper plate was more frequent at High Temperature (HT) than at RT. Additionally, Total Acid Number (TAN) values of the biodiesel were found to be increased at HT as compared to at RT [65]. Other researchers, Haseeb et al., conducted a similar study by using different blends of biodiesel for copper and leaded bronze by using the immersion test at room temperature for 2640 h. The blends which were used were B0, B50 and B100. The immersion tests were performed at 60 °C for 840 h for blends B0, B100 and B100 (oxidized). The corrosion rates of leaded bronze and copper in B100 were 0.018 and 0.042 mpy, respectively, at room temperature, while at 60 °C, the CRs of bronze and copper for B100 were 0.023 and 0.053 mpy, respectively. The corrosion rate for copper was higher than that of leaded bronze [55]. The following graphs in Figure 2 show the CR of copper and leaded bronze at RT and 60 °C.

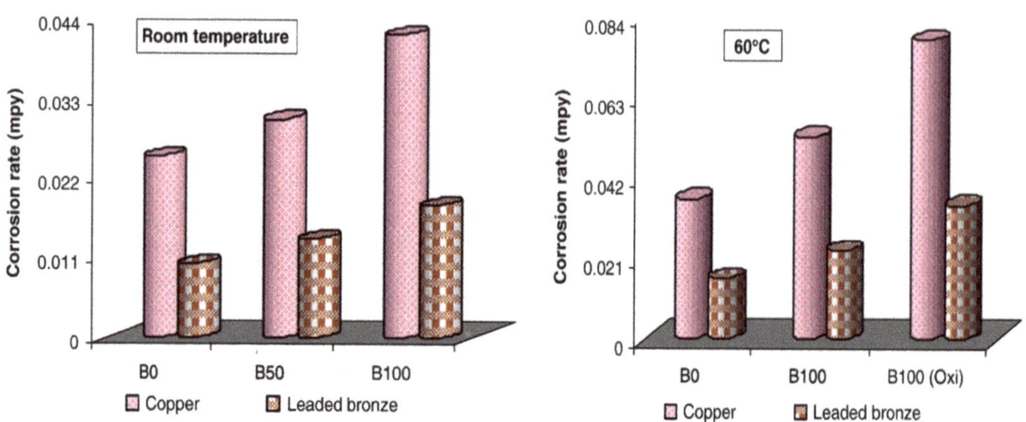

Figure 2. Corrosion Rates (CRs) of leaded bronze and copper at Room Temperature (RT) and 60 °C [55].

Another study was conducted by researchers for corrosion properties of Copper (Cu), Brass (BS), Aluminum (Al) and Cast Iron (CI) when exposed to palm biodiesel and diesel. The tests performed were of immersion type for B0 and B100 blends. These tests were performed for 2880 h at room temperature. The CRs of Cu, BS, Al and CI were 0.38278, 0.209898, 0.173055 and 0.112232 mpy, respectively. The TAN value of as-received biodiesel was 0.35 mg KOH/g. At the end of test, the TAN values of biodiesel increased to 2.57, 2.29, 1.68 and 1.69 mg KOH/g in the case of Cu, BS, Al and CI, respectively [64].

The same researcher performed another study on similar materials; however, this time it was not carried out at room temperature. This experiment was conducted at 80 °C and the duration was reduced this time to 1200 h. It was observed that the TAN number of the fuel was increased and it was higher than the limit according to ASTM D6751. It was observed that Cu and Al were more corrosive in biodiesel. At 1200 h and 80 °C, the corrosion rates for stainless steel (SS), Al and Cu in Palm biodiesel were 0.015, 0.202 and 0.586 mpy, respectively [33]. Figure 3 shows the corrosion rates of all materials.

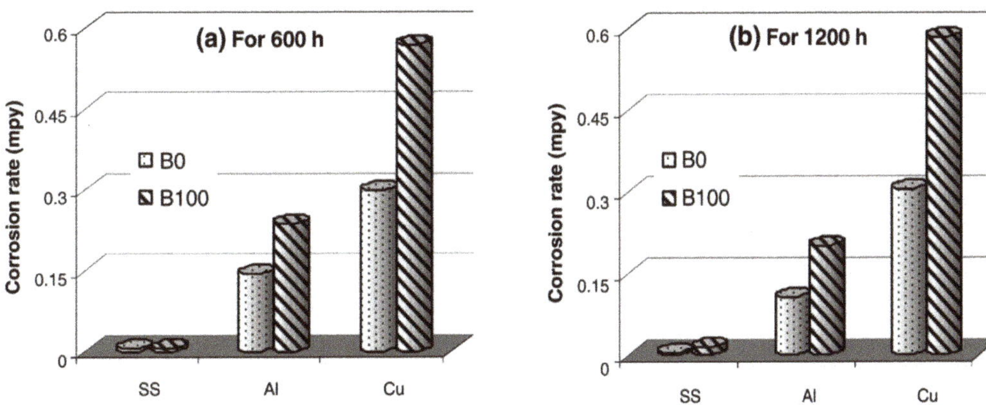

Figure 3. Corrosion rates of stainless steel (SS), aluminum (Al) and copper (Cu) [33].

From the above studies, it was observed that Cu showed higher corrosion rates followed by BS, Al and CI. SS showed a minimum corrosion rate in palm biodiesel. Moreover, when temperature and duration of immersion were increased the corrosion rate was increased and the values for corrosion rate can be seen in Table 1 as well.

Some studies were made by researchers to assess the corrosion behavior on carbon steels. Thangavelu et al. studied corrosion behavior of BDE blends B0 and B20D70E10 with carbon steel by immersion tests. The tests were performed at RT and 60 °C for 400 and 800 h. CR of B20D70E10 at RT and at 60 °C was more than the rate of B0 blend under the same conditions. At RT, it was 0.1817 mpy and at 60 °C, it was 0.2612 mpy for B20D70E10 and for B0 these values were 0.0523 mpy and 0.115 mpy. It was noticed that the TAN in B20D70E10 was exceeded by the limit. Initially it was 0.25 mg KOH/g for as-received biodiesel and after immersion it was increased to 1.15 mg KOH/g at room temperature and 1.59 mg KOH/g at 60 °C [42]. The surface morphology of carbon steel is shown in Figure 4.

58

Table 1. Summary of corrosion studies using palm oil-based biodiesel.

Sr. No.	Test Type	Material	Biofuel Blends	Temperature	Time	Characterization	TAN (mg KOH/g)	Corrosion Rate
1.	Immersion Test	Pure Copper	45% biodiesel, 35% diesel and 20% bioethanol (B45D35E20)	Room temperature 50 °C	408 h	OM, FTIR	Increased at high temperature	0.277 mpy [65] 0.327 mpy
2.	Immersion Test	Copper, Leaded Bronze	B0, B50, B100 B0, B100, B100(oxidized)	Room temperature 60 °C	2640 h 840 h	TAN analyzer, FTIR, MOA	-	0.018 mpy (Bronze) and 0.042 mpy (Copper) [55] 0.023 mpy (Bronze) and 0.053 mpy (Copper)
3.	Immersion Test	Copper, Brass, Aluminum, Cast Iron	B0, B100	Room temperature	2880 h	SEM/EDS, XRD	0.35 (as-received), 2.57 (Copper), 2.29 (Brass), 1.68 (Aluminum), 1.69 (Cast Iron)	0.38278 mpy (Copper), 0.209898 mpy (Brass), 0.173055 mpy (Aluminum) and 0.112232 mpy (Cast Iron) [64]
4.	Immersion Test	Copper, Stainless Steel, Aluminum	B0, B100	80 °C	1200 h	OM, SEM, EDS	Increased to limit according to ASTM D6751	0.015 mpy (Stainless Steel), 0.202 mpy (Aluminum) and 0.586 mpy (Copper) [33]
5.	Immersion Test	Carbon Steel	B0, B20D70E10	Room temperature 60 °C	800 h 400 h	OM, FTIR	0.25 (as-received) 1.15 (Room temperature) 1.59 (60 °C)	0.1817 mpy (B20D70E10) and 0.0523 (B0) [42] 0.2612 mpy (B20D70E10) and 0.115 mpy (B0)
6.	Immersion Test	Carbon Steel ASTM 1045	-	27 °C, 50 °C, 80 °C	30, 60, 120 days	SEM/EDS, XRD, FTIR	Increased with the increase in temperature and time	With the rise in temperature and exposure time the CRs of Mild Steel increased in both fuels [66]

Table 1. *Cont.*

Sr. No.	Test Type	Material	Biofuel Blends	Temperature	Time	Characterization	TAN (mg KOH/g)	Corrosion Rate
7.	Immersion Test	Mild Steel	B0, B50, B100	Room temperature, 50 °C, 80 °C	1200 h	OM, SEM, EDS, XRD	Increased to limit according to ASTM D6751	At RT, the CR of MS in biodiesel and diesel was 0.052 mpy and 0.046 mpy, respectively. While at 80 °C the CR was 0.059 mpy and 0.05 mpy, respectively [67]
8.	Immersion Test	Aluminum, Magnesium	B100	Room temperature	720 h / 1440 h	SEM/EDS, XRD, FTIR	0.27 (as-received) 0.92 (Aluminum) 0.87 (Magnesium)	0.1230 mpy (Aluminum) and 3.0910 mpy (Magnesium) [68] / 0.0527 mpy (Aluminum) and 2.6563 mpy (Magnesium)
9.	Immersion Test	Copper, Mild Steel	-	Room temperature	60 days	SEM/EDS, AFM	-	CR of Copper was decreased effectively by using Benzotriazole than that of Mild Steel. In the presence of Adenine, the CR of Copper decreased from 0.7495 to 0.2512 mg/cm^2 [69]

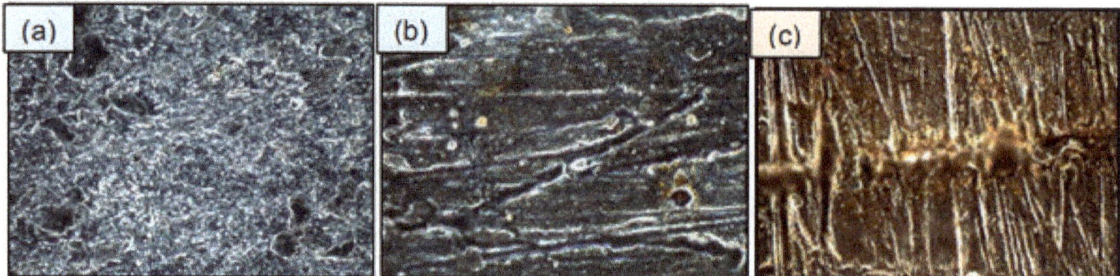

Figure 4. Surface morphology (**a**) B0 at RT; (**b**) B20D70E10 at RT; (**c**) B20D70E10 at 60 °C [42].

Another study was conducted by Jin et al. on ASTM 1045 MS for investigation of its corrosion behavior when immersed in palm biodiesel at 27, 50 and 80 °C for 30, 60 and 120 days, and these results were then compared with the commercial diesel. With the rise in temperature and exposure time the CRs of MS increased in both fuels. It was observed that corrosion rates obtained with commercial diesel were lower than those of palm biodiesel. The TAN values of biodiesel and diesel increase with the increase in exposure time and temperature [66].

B0, B50 and B100 blends of MS were again investigated for corrosion behavior by Fazal et al. at room temperature, 50 and 80 °C by static immersion tests for 1200 h. It was found that petroleum diesel is less corrosive than the biodiesel. At RT, CRs of MS in biodiesel and diesel were 0.052 and 0.046 mpy, respectively. At 80 °C, the corrosion rates were 0.059 and 0.05 mpy, respectively. The water content was increased by increasing immersion temperature. Additionally, it was observed that the TAN of fuel was increased and it was higher than the limit according to ASTM D6751 [67].

Studies on carbon steels showed that CR increases by increasing the concentration of biodiesel in diesel for carbon steel and mild carbon steel. With the increase in temperature, the CR had increased for all carbon steels presented above. Additionally, it was noticed that the duration of immersion had affected CR and it was increased with the increase in duration. Petroleum diesel was found to be the least corrosive. TAN values of the biodiesel increased. The values of CR can be seen in Table 1. It concludes that the values of CR are directly proportional to the duration of immersion, temperature and blend percentage.

Another work conducted by Chew et al. assessed the corrosion behavior of aluminum and Magnesium (Mg) for 720 and 1440 h by immersion test. The corrosion rates of Al were 0.1230 and 0.0527 mpy at 720 and 1440 h, respectively. Similarly, Mg showed rates of 3.0910 and 2.6563 mpy at 720 and 1440 h, respectively. Mg exhibits higher corrosion rates as compared to Al was due to the higher reactivity of magnesium. It was noted that Mg was less noble as compared to Mg in galvanic series. The corrosion rates of both materials decreased with the increase in duration, as shown in Table 1. The TAN values of both materials after exposure to biodiesel enhanced from 0.27 mg KOH/g of as-received biodiesel to 0.87 and 0.92 mg KOH/g of biodiesel in which Mg and Al were immersed, respectively [68]. Figure 5 shows the corrosion rates of both materials.

Summaries of these studies are shown in Table 1 below.

4.2. Corrosion Studies on Jatropha Oil-Based Biodiesel

There are a few studies on Jatropha oil-based biodiesels to assess corrosion rates of automotive materials. A study was completed by Dharma et al. where he used 50% jatropha curcas and 50% ceiba pentandra (J50C50) biodiesels to assess corrosion behavior of MS by static immersion tests at room temperature for 400 and 800 h for blends B0, B10, B20, B30, B40 and B50. The corrosion rates at all mixtures were higher at all blends at 400 h and these were 0.0018, 0.0011, 0.0198, 0.0199, 0.0222 and 0.0289 mm/year for B0, B10, B20,

B30, B40 and B50, respectively, as shown below in Table 2. The corrosion rate for B50 was 15 times higher than that of diesel fuel. It can be noticed that the weight loss is not linear with immersion time and it tends to slow down as the duration increases. Therefore, it can be noticed that for B20, B30 and B50, corrosion rates decreased at 800 h and were 0.01176, 0.01546 and 0.02524 mm/year, respectively, as shown in Table 2. The acid value of the fuel was higher, although it was still in range as compared to ASTM D 6751 standards [41].

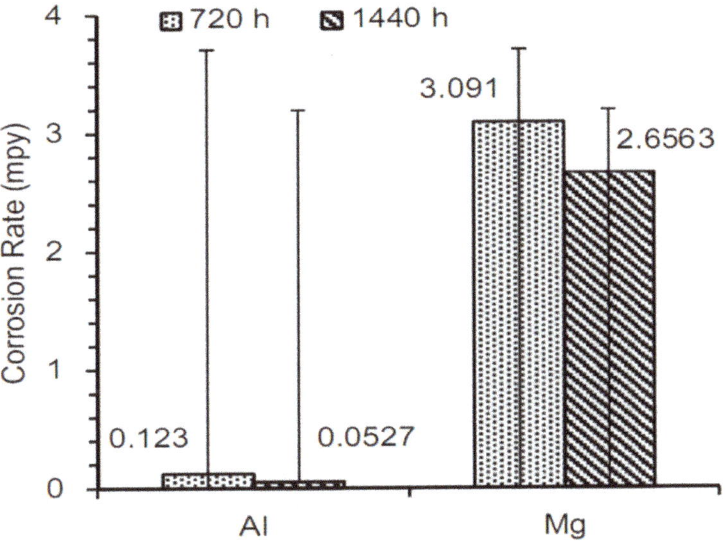

Figure 5. Corrosion rates of magnesium and aluminum at 720 and 1440 h [68].

Another study was conducted by Akhabue et al. on Al and MCS in Jatropha oil-based biodiesel by static immersion tests for blends B0, B50 and B100 at RT for 18 weeks (3 weeks interval). It was observed that in the case of MCS, there was an increase in CR up to the 12th week. The CR of B50 remained same between the 12th and 15th weeks. However, it was reduced for B0 and B100 after the 12th week. At the end of the experiment, the CRs of MCS in B0, B50 and B100 were 0.0011, 0.0022 and 0.0026 mpy, respectively, as shown in Table 2. In the case of Al, CRs were found to be increased up until the 12th week. The maximum corrosion rates for B50 and B100 were obtained in the 15th week, reaching 0.0099 and 0.016 mpy, respectively, while for B0 the maximum value was obtained in the 12th week. However, there was a decreased CR for B0 in the 15th week. B50 showed a decrease in rate between the 9th and 12th weeks. The same CRs were found to be increased from the 12th week in B50 and B100. If compared, the corrosion rates of Al were lower than those of MCS under the same conditions as shown in Table 2. The TAN values of as-received biodiesel were 0.41 and 0.52 mg KOH/g for B100 and B50, respectively, while after immersion these values were 3.53 and 1.54 mg KOH/g for B100 and B50, respectively, in the case of MCS, and 2.81 and 1.51 mg KOH/g for B100 and B50, respectively, in the case of Al [62].

Table 2. Summary of corrosion studies using jatropha oil-based biodiesel.

Sr. No.	Test Type	Material	Biofuel Blends	Temperature	Time	Characterization	TAN (mg KOH/g)	Corrosion Rate
1.	Immersion Test	Piston metal, Piston liner	B100	Room temperature	300 days	-	0.38 (as-received), 19.54 (piston liner), 14.48 (piston metal)	Salvadora is the most corrosive biodiesel followed by jatropha curcas, mahua and karanja [70]
2.	Immersion Test	Mild Steel	B0, B10, B20, B30, B40, B50	Room temperature	400 h, 800 h	SEM	Higher however in range as compared to ASTM D 6751 standards	CRs were higher at 400 h and were 0.0018 (B0), 0.0011 (B10), 0.0198 (B20), 0.0199 (B30), 0.0222 (B40) and 0.0289 (B50) mm/year. For B20, B30 and B50, CRs decreased to 0.01176, 0.01546 and 0.02524 mm/year at 800 h [41]
3.	Immersion Test	Aluminum, Mild Carbon Steel	B0, B50, B100	Room temperature	18 weeks (3 weeks interval)	-	0.41 and 0.52 (as-received B100 and B50), 3.53 and 1.54 (MCS B100 and B50), 2.81 and 1.51 (Al B100 and B50)	The CRs of Aluminum were lower than those of MCS. The CRs of MCS in B0, B50 and B100 were 0.0011, 0.0022 and 0.0026 mpy, respectively. The CRs of Al in B50 and B100 were 0.0099 and 0.016 mpy in the 15th week [62]

In another study, researchers used biodiesels of Indian origin (salvadora oleoides, madhuca indica, pongamia glabra and jatropha curcas) for the corrosion behavior of materials by static immersion method for 300 days at room temperature on piston metal and piston liner parts. In the case of piston liner, there were clear weight losses observed in all samples of biodiesel. Weight loss was slightly higher in the case of jatropha curcas biodiesel. In the case of piston metal, the corrosion in mahua and karanja oisl was comparable to that of corrosion in petro diesel, while in that of jatropha curcas was slightly higher. The weight loss in salvadora was 10 times higher than that obtained in jatropha curcas oil as shown in Table 2. The TAN values for salvadora oleoides, madhuca indica, pongamia glabra and jatropha curcas were 0.45, 0.32, 0.42 and 0.38 mg KOH/g, respectively, before immersion, while after immersion these values were 2.51, 19.72, 17.52 and 19.54 mg KOH/g, respectively, for piston liner and 2.38, 11.3, 14.39 and 14.48 mg KOH/g, respectively, for piston metal [70].

Hence, it was clear from the above studies that jatropha biodiesel is also corrosive for automotive materials and the corrosiveness increases by increasing the concentration of biodiesel. However, it did not show any particular trend with the duration of immersion and in some cases the CR was decreased by increasing the duration. Furthermore, it was noticed that Al did not show any significant corrosion in this biodiesel and CR of MCS was higher as compared to Al as shown in Table 2. However, Jatropha biodiesel showed the same trends with the TAN values and these were increased by the use of biodiesel.

Summaries of these studies have been shown in Table 2 below.

4.3. Corrosion Studies on Rapeseed Oil-Based Biodiesel

Various researchers conducted studies to find the corrosion rate of automotive materials by using rapeseed oil-based biodiesel. A study was carried out by Sterpu et al. in which he investigated corrosion of Carbon Steel (CS) in corn, rapeseed and sunflower biodiesels by immersion tests at RT for 1176 h. The CRs of CS in corn, rapeseed and sunflower biodiesel were 0.001164, 0.000760 and 0.000855 mm/year, respectively, as shown in a graph in Figure 6. Likewise, the TAN of biodiesel was increased [63].

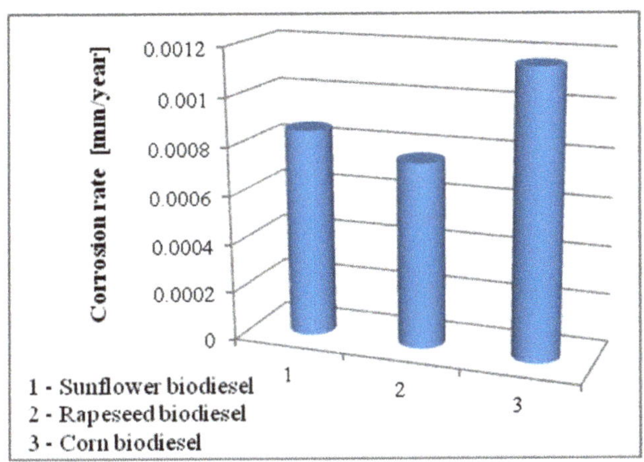

Figure 6. CRs of carbon steel in three different biodiesels [63].

In another work, Hu et al. used the B20 blend of rapeseed biodiesel to study the corrosion behaviors of SS, Al, MCS and Cu by immersion tests at 43 °C for 2 months. The corrosion of Al and SS was much lower than the corrosion of MCS and Cu in biodiesel. The corrosion rates of Cu and MCS were 0.02334 and 0.01819 mm/year in biodiesel while 0.0037 and 0.0015 mm/year for diesel, respectively, as shown in Figure 7 [71].

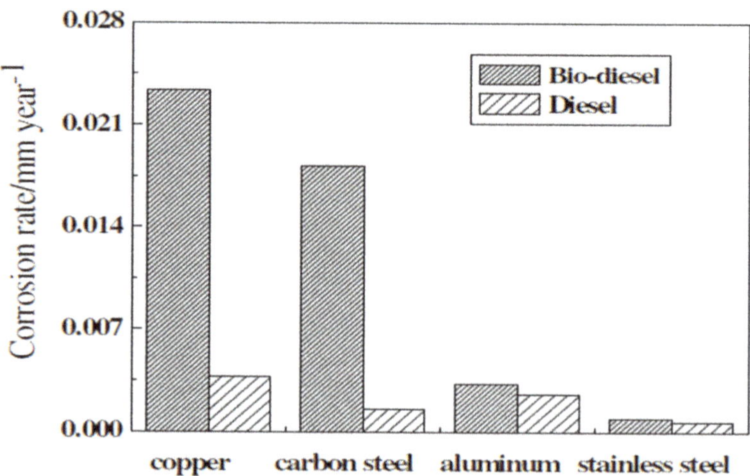

Figure 7. CRs of metals in diesel and biodiesel [71].

Additionally, B0, B50, B75 and B100 blends of rapeseed methyl ester with ultra-low sulfur diesel fuel were used by Norouzi et al. to study the corrosion of AW 6060 aluminum alloy and E-Cu57 copper by immersion tests 600 h at 80 °C. They observed that the increase in biodiesel concentration enhanced the corrosion of the biodiesel, as shown in Figure 8. Additionally, the increased temperature increased the corrosiveness of biodiesel. The TAN of biodiesel was initially 0.315 mg KOH/g and after immersion, it was more than the limit according to ASTM D6751, which is 0.8 mg KOH/g for biodiesel. The color of the biodiesel changed in both materials due to the change in the composition of the biodiesel [72].

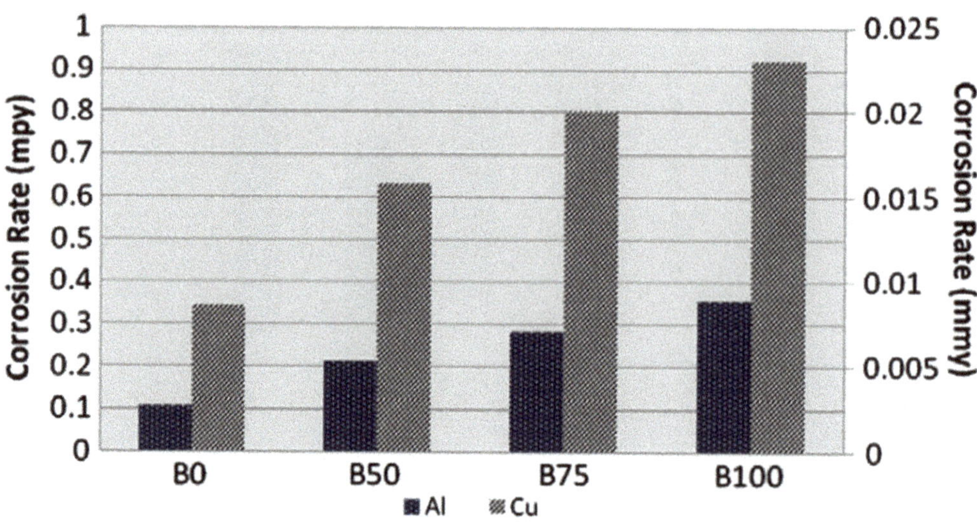

Figure 8. CRs of Cu and Al in fuel blends [72].

In the case of rapeseed biodiesel, it can be concluded that this biodiesel showed similar trends as compared to other biodiesels. The increased concentration of biodiesel and increased temperature increased the corrosion rates of metals, as shown in Table 3. Furthermore, TAN values of the biodiesel were found to be increased.

Summaries of these studies have been shown in Table 3 below.

4.4. Corrosion Studies on Sunflower Oil-Based Biodiesel

Some of the studies were conducted to find the corrosion rates of automotive materials by using sunflower oil-based biodiesels. In their study, Samuel et al. investigated corrosion properties of brass in B10, B20 and B40 blends of waste sunflower oil biodiesel (WSOB) by immersion tests at room temperature for 240, 480, 720 and 960 h. By increasing duration and the biodiesel fraction, the corrosion of the BS increased. The TAN of the as-received fuel was 0.297 mg KOH/g, which increased to 0.35, 0.4, 0.47, 0.73 and 1.95 mg KOH/g for B0, B10, B20, B40 and B100 blends, respectively. This was because of the acid component variation [73]. A similar study on B0, B20 and B100 was performed by Cursaru et al. to assess corrosion rates of ferrous alloy, Cu and Al using the static immersion test at RT and 60 °C for 3000 h. The corrosion rate of each metal increased by increasing the fraction of biodiesel in the blend. At room temperature, the corrosion rates of Cu, Al and MCS were 0.323615, 0.162201 and 0.170124 mpy, respectively, as shown in Table 4 [74]. Similarly, at 60 °C, these results were 0.640758, 0.316292 and 0.336845 mpy, respectively.

Another study was presented by Cursaru et al. and this time he used the electrochemical method for corrosion rate calculation and the materials used were changed to monel steel, stainless steel and mild steel. The biodiesel used in this study was in percentages of 1, 5 and 10 for 3 h at 50 °C. The TAN of as-received fuel was 0.12 mg KOH/g which was increased to 0.18, 0.2, 0.21 and 0.3 mg KOH/g for B1, B5, B10 and B100 blends, respectively. The monel steel was the least corroded as compared to stainless steel and mild steel. The corrosion rate of monel steel was 0.000045 mm/year, stainless steel was 0.000421 mm/year and mild carbon steel was 0.000514 mm/year, as shown in Table 4 [75].

Additionally, the trends of sunflower biodiesel were similar in terms of corrosion rate to other biodiesels as the CRs of materials in it were increased by the increase in the concentration of biodiesel and duration. The corrosion rate of diesel was lower as compared to biodiesel. Furthermore, an increased temperature increased the corrosion, as shown below in Table 4. TAN of the biodiesel was increased as well.

Summaries of these studies have been shown in Table 4 below.

4.5. Corrosion Studies on Biodiesel Obtained from Different Feedstocks

Many other studies to assess the corrosion rates of materials were presented by several researchers by using biodiesel obtained from the different feedstocks. Diaz-Ballote et al. used canola biodiesel to assess the corrosion rate of Al by an electrochemical technique. It was observed that the corrosion of Al can be used as an indicator to assess the purity of the biodiesel as Al shows corrosiveness in biodiesel [76]. Another work by Ononiwo et al. investigated MS for corrosion by immersion tests using ghee butter-based biodiesel at room temperature and different temperatures. The weight losses in mineral diesel and biodiesel samples were quite close. MS showed a similar response in all the media studied. Mineral diesel showed lower weight loss than samples 1 and 2, as shown in Table 5 [4]. The corrosion rate was increased by the increase in the temperature.

Table 3. Summary of corrosion studies using rapeseed biodiesel.

Sr. No.	Test Type	Material	Biofuel Blends	Temperature	Time	Characterization	TAN (mg KOH/g)	Corrosion Rate
1.	Immersion Test	Carbon Steel	-	Room temperature	1176 h	-	Increased	Corrosion rates in corn, rapeseed and sunflower biodiesels were 0.001164, 0.000760 and 0.000855 mm/year, respectively [63]
2.	Immersion Test	Copper, Aluminum, Stainless Steel, Mild Carbon Steel	B20	43 °C	2 months	SEM/EDS, XPS, AAS	-	The corrosion rates of copper and mild carbon steel were 0.02334 and 0.01819 mm/year in biodiesel while 0.0037 and 0.0015 mm/year for diesel, respectively [71]
3.	Immersion Test	AW 6060 Aluminum alloy, E-Cu57 Copper	B0, B50, B75, B100	80 °C	600 h	SEM/EDS	More than the ASTM D6751 limit of 0.8	The increase in biodiesel concentration and the temperature increased the corrosion rate of the biodiesel [72]

Table 4. Summary of corrosion studies using sunflower biodiesel.

Sr. No.	Test Type	Material	Biofuel Blends	Temperature	Time	Characterization	TAN (mg KOH/g)	Corrosion Rate
1.	Immersion Test	Brass	B0, B10, B20, B40	Room temperature	240, 480, 720, 960 h	JCM 100 mini SEM	0.297 (as-received), 0.35 (B0), 0.4 (B10), 0.47 (B20), 0.73 (B40), 1.95 (B100)	Increased duration and biodiesel fraction increased the corrosion rate [73]
2.	Immersion Test	Aluminum, Copper, Mild Carbon Steel	B0, B20, B100	Room temperature, 60 °C	3000 h	SEM/EDS, XRD	-	At RT, the CRs of copper, aluminum and mild carbon steel were 0.323615, 0.162201 and 0.170124 mpy, respectively, and at 60 °C, these results were 0.640758, 0.316292 and 0.336845 mpy, respectively [74]
3.	Electrochemical Method	Monel Steel, Stainless Steel, Mild Steel	1, 5 and 10 percent	50 °C	3 h	-	0.12 (as-received), 0.18 (B1), 0.2 (B5), 0.21 (B10), 0.3 (B100)	The CR of monel steel was 0.000045 mm/year, stainless steel was 0.000421 mm/year and mild carbon steel was 0.000514 mm/year. Corrosion of these materials was lowest in pure diesel than in biodiesel. [75]

Table 5. Summary of corrosion studies on biodiesel of different feedstock.

Sr. No.	Test Type	Biodiesel	Material	Biofuel Blends	Temperature	Time	Characterization	Corrosion Rate/Findings
1.	Electrochemical	Canola	Aluminum	-	-	-	-	The corrosion of Aluminum can be used as an indicator to assess the purity of the biodiesel [76]
2.	Immersion Test	Ghee Butter	Mild Steel	-	Room temperature, Different temperatures	-	-	Biodiesel samples were more corrosive than mineral diesel samples. The CR was increased with the increase in the temperature [4]
3.	Immersion Test	Rice Husk	Austenite Stainless Steel, Brass, Mild Steel, Aluminum	B10, B30	Room temperature, 50 °C, 70 °C	6, 12, 24, 48, 72, 120 h	XPS	Brass exhibited very slight weight loss, Aluminum exhibited little more than Brass and Mild Steel exhibited more weight loss than both Aluminum and Brass. At elevated temperatures, the weight loss was enhanced for the above-mentioned materials. Stainless Steel did not show any weight loss at any conditions [77]

Table 5. *Cont.*

Sr. No.	Test Type	Biodiesel	Material	Biofuel Blends	Temperature	Time	Characterization	Corrosion Rate/Findings
4.	Electrochemical	Soybean	Duplex 2205, Sea Curve, AISI 304l	-	Room temperature	20 h	-	The best resistance towards corrosion was shown by Duplex 2205 while Sea Curve steel showed the least corrosion resistance [78]
5.	Immersion Test	Commercial	Copper, Brass	-	55 °C	5 days	Raman Vibrational Spectroscopy	The rate of corrosion was slightly higher in the case of incidence light. At higher temperature, corrosion rates have decreased. The CRs for Brass are always more than those of Copper [1]
6.	Immersion Test	Poultry Fat	316 Stainless Steel, Grey Cast Iron, Copper, Admiralty Brass, Carbon Steel	B20, B80	38 °C	10 months	Digital Photography, photo Microscopy	316 Stainless Steel and Carbon Steel had no visible corrosion. Copper showed huge corrosiveness in biodiesel. B20 had a slightly lower corrosion rate than that of B80. In the case of Brass, similar trends to that of Copper were observed however these were to a lesser extent [79]

69

B10 and B30 blends of rice husk bio-oil were used by Lu et al. to assess the corrosion properties of four materials—SS, MS, BS and Al—by immersion tests at RT, 50 and 70 °C for 6, 12, 24, 48, 72 and 120 h. BS exhibited very slight weight loss, Al exhibited little more than BS and MS exhibited more weight loss than both Al and BS. SS did not show any weight loss at any conditions [77]. The corrosion resistance of Duplex 2205, Sea Curve and AISI 304l stainless steel were additionally studied by Roman et al. in soybean biodiesel by an electrochemical technique at room temperature for 20 h. It was observed that all the steels showed good performances towards the corrosion resistance. The most resistive towards corrosion was Duplex 2205, while Sea Curve steel showed the least corrosion resistance, as shown in Table 5. Therefore, it was concluded that all these steels can be used in applications that need to be in contact with soybean oil-based biodiesel [78].

A study was conducted by Aquino et al. to investigate CRs of Cu and BS when immersed in commercial biodiesel in the presence of natural light and temperature for 5 days at 55 °C. The rate of corrosion was slightly higher in the case of incidence light. The corrosion rates for BS were always higher than those of Cu. It was found that the corrosion rates decreased at high temperatures, as shown in Table 5, which is contradictory to expected results because normally increases in temperature increase corrosion rates because of the increase in the reaction rate. The reason behind this result could be the elimination of the absorbed oxygen at high temperatures in the biodiesel [1]. Another study was performed by Geller et al. to investigate corrosion behaviors of B20 and B80 blends of poultry fat biodiesel on stainless steel 316, grey cast iron, copper, admiralty brass and carbon steel. Static immersion tests were performed for 10 months at 38 °C. Stainless steel 316 and carbon steel showed no weight loss or visible corrosion. Copper showed huge corrosiveness in biodiesel. B20 had slightly lower weight loss 0.71% than that of B80 0.74% [79].

As shown in Table 5, it is clear that all biodiesels showed increased corrosion rates for all materials by the increase in the concentration of biodiesel. There were few deviations shown by different biodiesels concerning duration and temperature; however, in most cases, the corrosion rate was increased by increasing the duration and temperature of immersion. Therefore, it is concluded that the CR of materials in biodiesel increases by increasing concentration, duration and temperature. Summaries of these studies have been shown in Table 5.

5. Corrosion Mechanism of Metals in Biodiesel

The X-ray Diffraction (XRD) results of Cu showed the formation of different compounds such as Cu_2O, CuO [80], $CuCO_3$ [81] and $Cu(OH)_2$ [82] on the surface of the base metal. The mechanism or reactions which took place during this process are stated below [83].

$$Cu_2O + 1/2O_2 \rightarrow 2CuO$$
$$Cu_2O + CO_2 + 1/2O_2 \rightarrow 2CuCO_3$$
$$CuO + CO_2 \rightarrow CuCO_3 \tag{2}$$
$$2Cu + O_2 + 2H_2O \rightarrow 2Cu(OH)_2$$
$$CuO + H_2 \rightarrow Cu(OH)_2$$

RCOO• radical [84] generation was thought to be the main cause of copper carbonate formation [85] through the decomposition of esters [86].

$$RCOOR_1 \rightarrow RCOO^* + R_1^* \tag{3}$$

$$2RCOO^* + Cu \rightarrow CuCO_3 + R—R + CO$$

$$2Cu(OH)_2 + CO_2 \rightarrow 2Cu(OH)_2 \cdot CuCO_3 + H_2O$$

As iron is the main component of steel, when exposed to biodiesel, it is more prone to corrosion. Therefore, XRD results showed the formation of some iron products such as

Fe_2O_3 [71], $Fe(OH)_2$ [87] and $FeCO_3$ [88]. The mechanism or reactions which took place during that process are stated below [89].

$$RCOOR_1 \rightarrow RCOO^* + R_1^* \tag{4}$$

$$4Fe + 3O_2 \rightarrow 2Fe_2O_3$$

$$2Fe + O_2 + 2H_2O \rightarrow 2Fe(OH)_2$$

$$4Fe(OH)_2 + O_2 \rightarrow 2Fe_2O_3. H_2O + 2H_2O$$

$$RCOOR_1 \rightarrow RCOO^* + R_1^*$$

$$2RCOO^* + Fe \rightarrow FeCO_3 + R\!-\!R + CO\,-$$

6. Characterization Techniques Used

Characterization techniques that have been used by researchers for corrosion testing include Scanning Electron Microscope/Energy Dispersive Spectroscopy (SEM/EDS), X-ray diffraction (XRD), Optical Microscopy (OM), X-ray Photoelectron Spectroscopy (XPS), Atomic Absorption Spectrometry (AAS) and Raman vibrational spectroscopy. SEM/EDS and XRD used by researchers concerning the biodiesel are stated below.

6.1. Characterization and Products of Corrosion Obtained by Palm Biodiesel

When palm biodiesel was used to find CRs of copper and leaded bronze, the oxide layer obtained at RT was copper carbonate ($CuCO_3$) and at 60 °C was of cupric oxide (CuO) [55].

When palm biodiesel was used to find the CR of CS ASTM 1045 at 27, 50 and 80 °C for 30, 60 and 120 days, SEM results showed that the effects of corrosion become severe as the exposure time and temperature were increased, as shown in the Figures 9 and 10. The attacked zones were not continuous and were scattered over the whole surface.

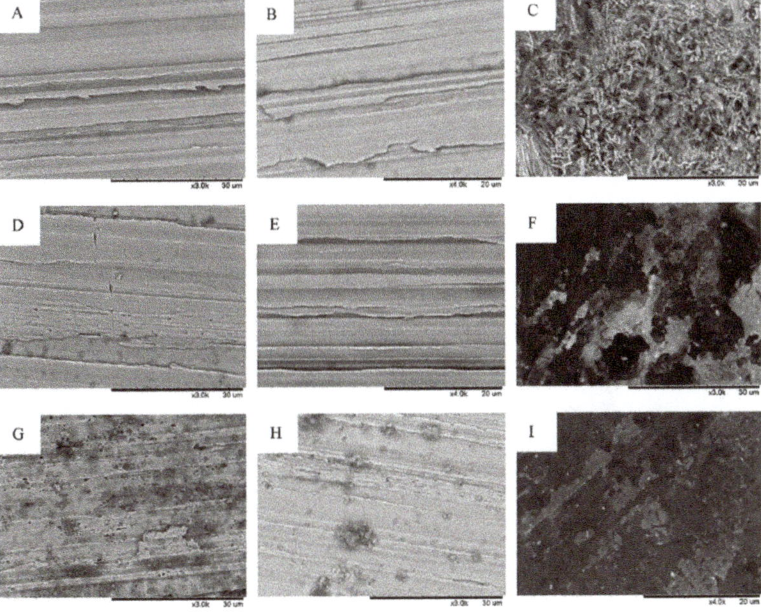

Figure 9. Surface morphology of ASTM 1045 MS in 0# diesel at different temperatures 27, 50 and 80 °C for different exposure periods: (**A**) 27 °C—30 d; (**B**) 50 °C—30 d; (**C**) 80 °C—30 d; (**D**) 27 °C—60 d; (**E**) 50 °C—60 d; (**F**) 80 °C—60 d; (**G**) 27 °C—120 d; (**H**) 50 °C—120 d; (**I**) 80 °C—120 d [66].

Figure 10. Surface morphology of ASTM 1045 MS in palm biodiesel at different temperatures 27, 50 and 80 °C for different exposure periods: (**A**) 27 °C—30 d; (**B**) 50 °C—30 d; (**C**) 80 °C—30 d; (**D**) 27 °C—60 d; (**E**) 50 °C—60 d; (**F**) 80 °C—60 d; (**G**) 27 °C—120 d; (**H**) 50 °C—120 d; (**I**) 80 °C—120 d [66].

XRD showed that the compounds formed on the surface of MS in biodiesel were Fe_2O_3, $FeO(OH)$, $Fe_2O_2CO_3$, FeO and $FeCO_3$. The products formed on the commercial diesel surface were FeO, $FeO(OH)$ and Fe_2O_3, as shown in Figure 11 [66].

When palm biodiesel was used to find CR of Al and Mg at RT for 720 and 1440 h, on the SEM image at 1000X magnification of the as-received materials, only grinding lines were visible. At 1440 h, again grinding lines were visible on the material surface of Al, therefore it did not achieve any significant change. However, the surface of Mg showed significant change and its surface was degraded because of the corrosion attack, as shown in Figure 12. XRD analysis did not show the formation of any compound on the surface of both materials [68].

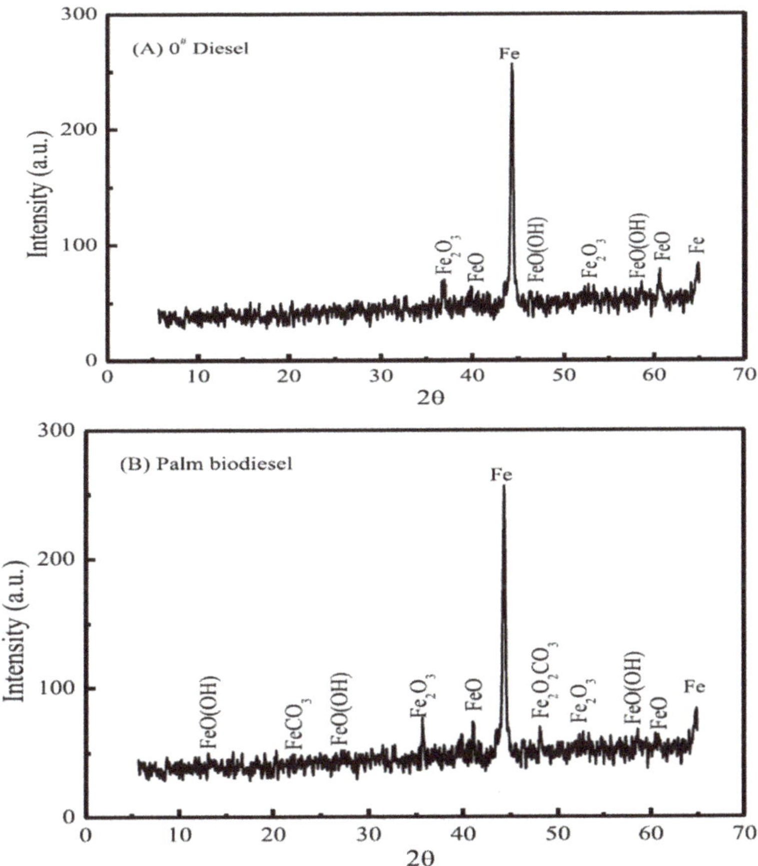

Figure 11. Corrosion products formed on MS at #0 diesel and biodiesel [66].

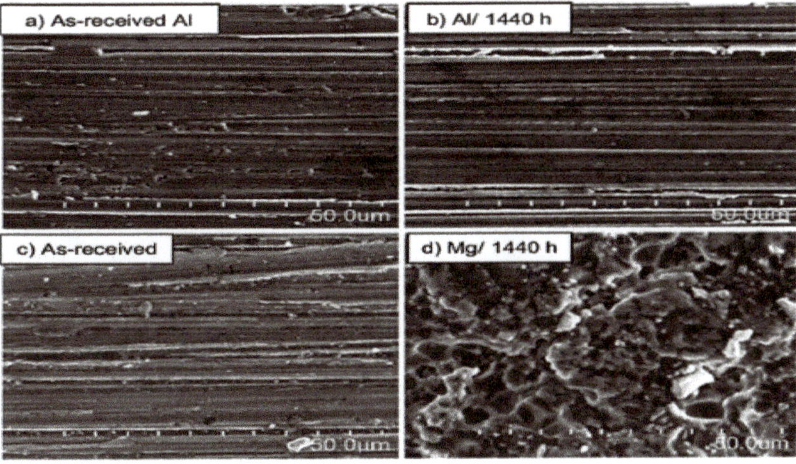

Figure 12. Scanning Electron Microscope (SEM) image of Mg and Al before and after immersion [68].

When palm biodiesel was used to find corrosion rates of MS at RT, 50 and 80 °C for 1200 h, the XRD results showed the presence of certain compounds on the surface. Biodiesel had Fe_2O_3, $Fe_2O_2CO_3$ and $Fe(OH)_3$ while diesel showed only Fe_2O_3 and $Fe(OH)_3$, as shown in Figure 13 [67]. Figure 14 shows the SEM images of samples before and after the immersion. It can be clearly seen that materials exhibit corrosion after the immersion. The zones where corrosion attacked are discontinued and scattered over the surface.

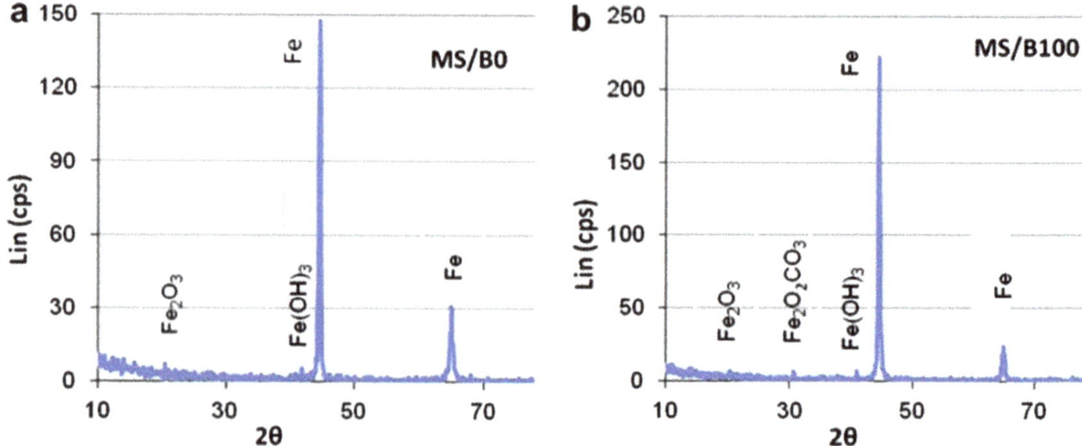

Figure 13. Corrosion products formed on MS upon exposure to diesel and biodiesel [67].

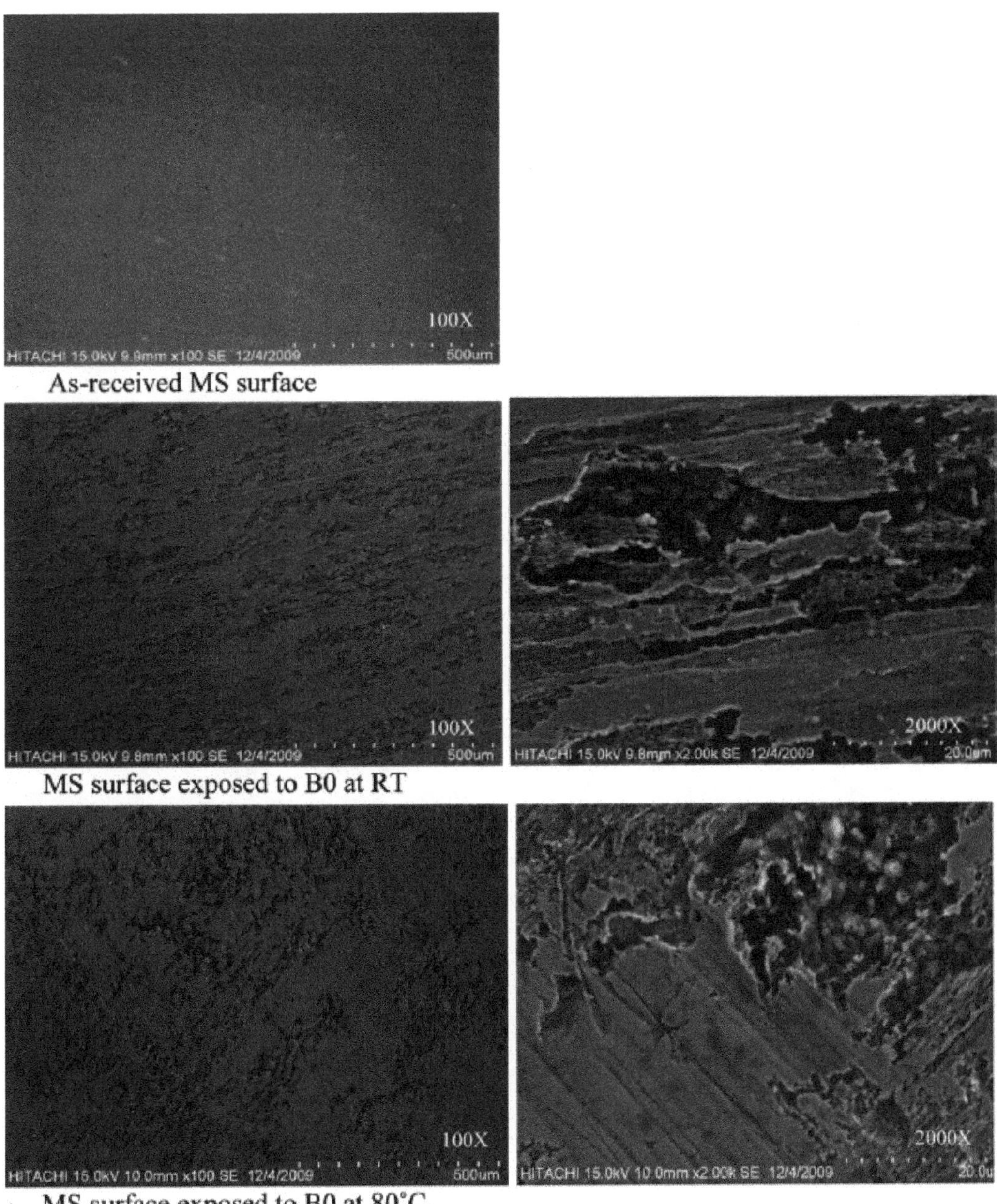

Figure 14. *Cont.*

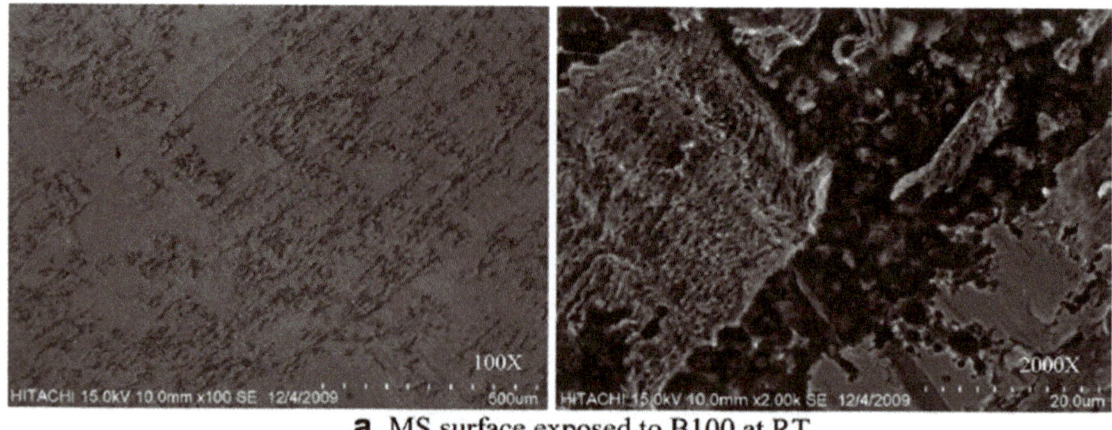

a MS surface exposed to B100 at RT

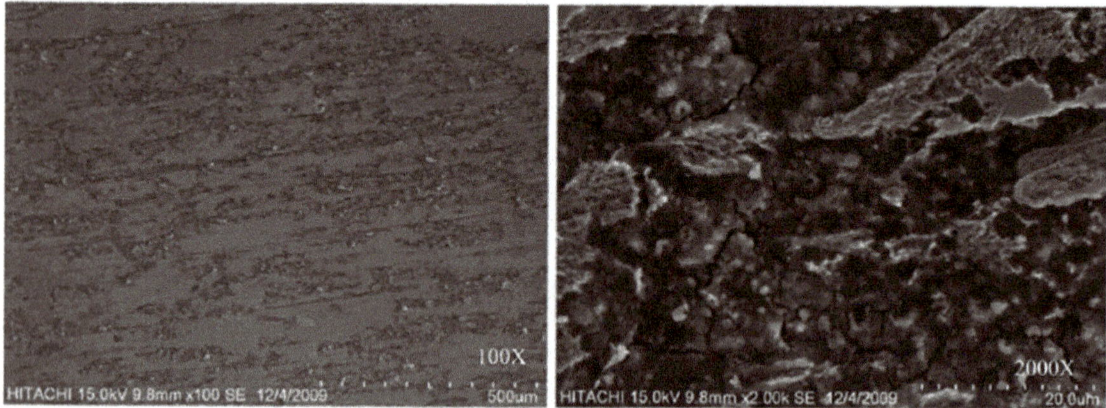

b MS surface exposed to B100 at 80°C

Figure 14. SEM images of as-received MS and on exposure to B0 (up) and B100 (down) at room temperature and 80 °C [67].

When palm biodiesel was used for Cu, BS, Al and CI at RT for 2880 h, the XRD results in Figure 15i,ii showed the formation of $CuCO_3$ in higher concentration along with CuO, $CuCO_3.Cu(OH)_2$, $Cu(OH)_2$ and Cu_2O. Moreover, in brass, small amounts of $CuCO_3$ seemed to form. In the case of aluminum, no compound was formed on its surface in biodiesel or diesel. SEM images in Figure 16 showed that the corrosion attack in biodiesel is more than that of diesel. Aluminum showed the lowest corrosion attack as compared to other materials [64].

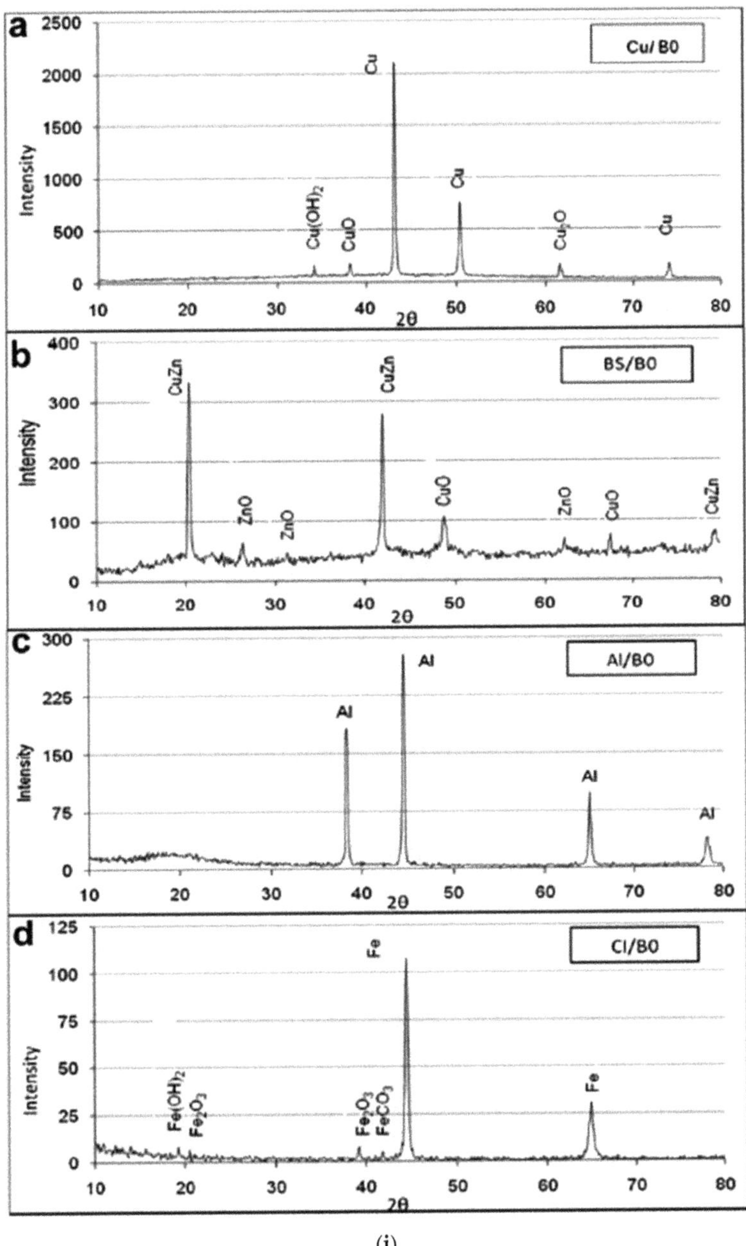

(i)

Figure 15. *Cont.*

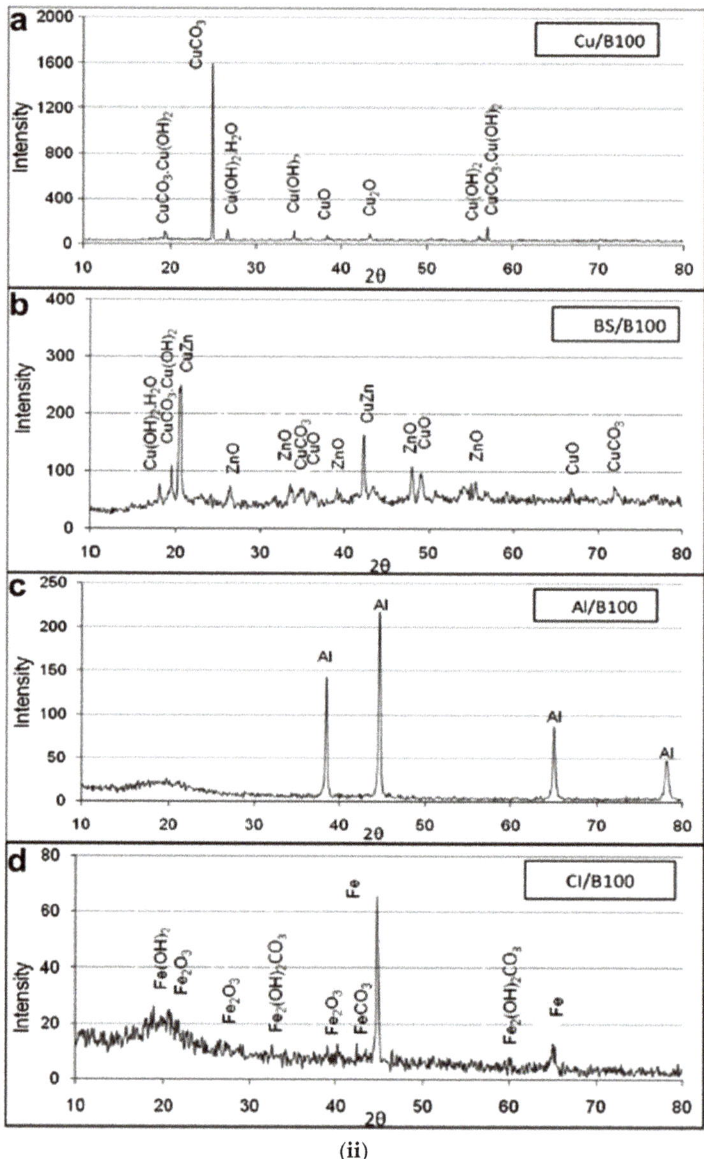

(ii)

Figure 15. (i) Products of corrosion formed on as-received materials and B0 and B100 blends [64]. (ii) Products of corrosion formed on as-received materials and B0 and B100 blends [64].

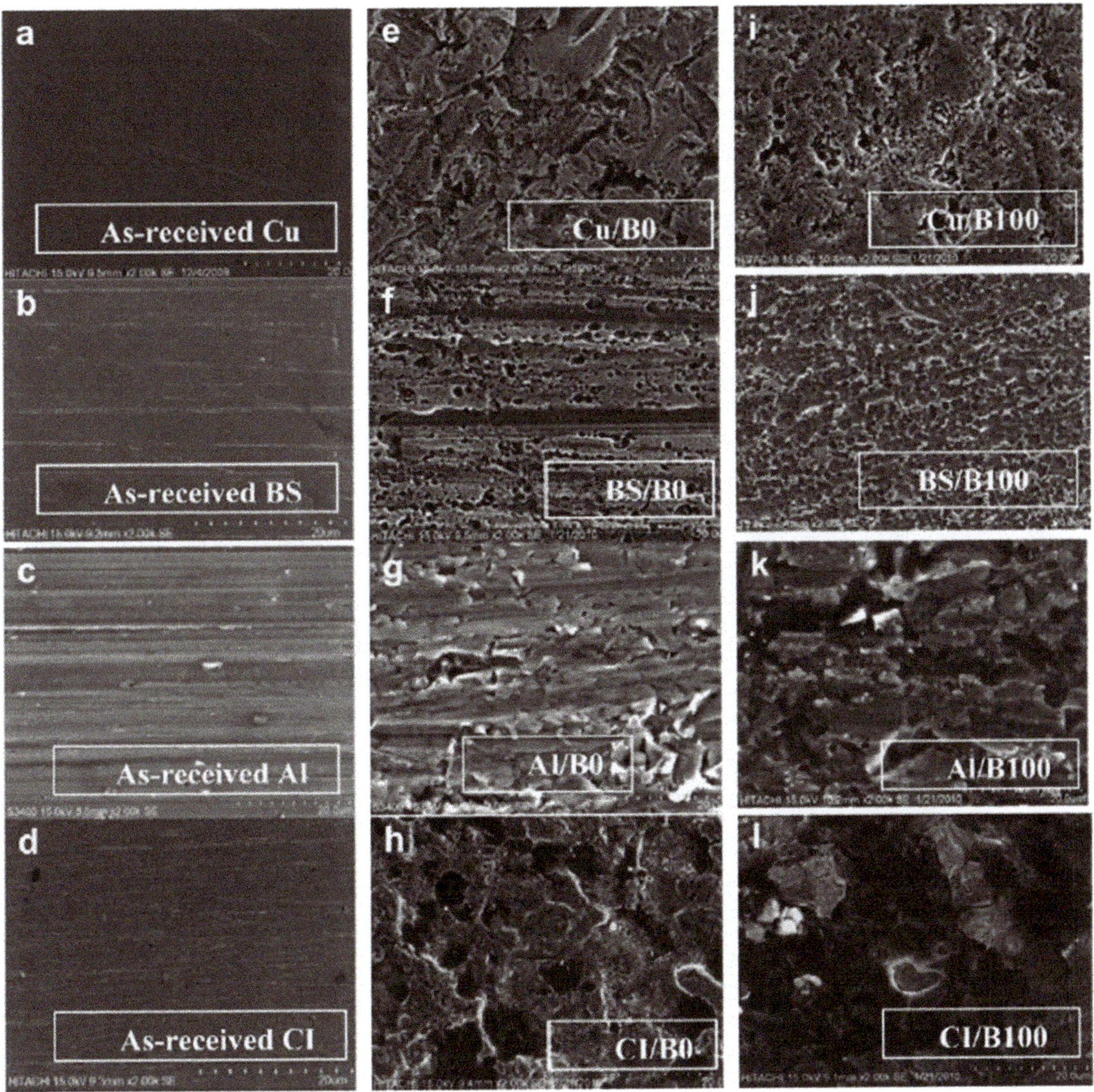

Figure 16. SEM images of as-received materials and when exposed to B0 and B100 blends [64].

6.2. Characterization and Products of Corrosion Obtained by Rapeseed Biodiesel

When the CRs of Cu, Al, SS and MCS were assessed by rapeseed biodiesel for two months at 43 °C, the SEM images of the materials showed that there were clear changes in the surface morphologies of all materials because of corrosion, except stainless steel, where only a small change was visible, as shown in Figure 17, which was because of the reaction with biodiesel. In the case of copper, a black layer covered its surface. The surface of aluminum was turned slightly dark. In the case of stainless steel, there were no changes on the surface. The XPS showed the formation of metal oxides including CuO, Cu_2O, Fe_2O_3 [71].

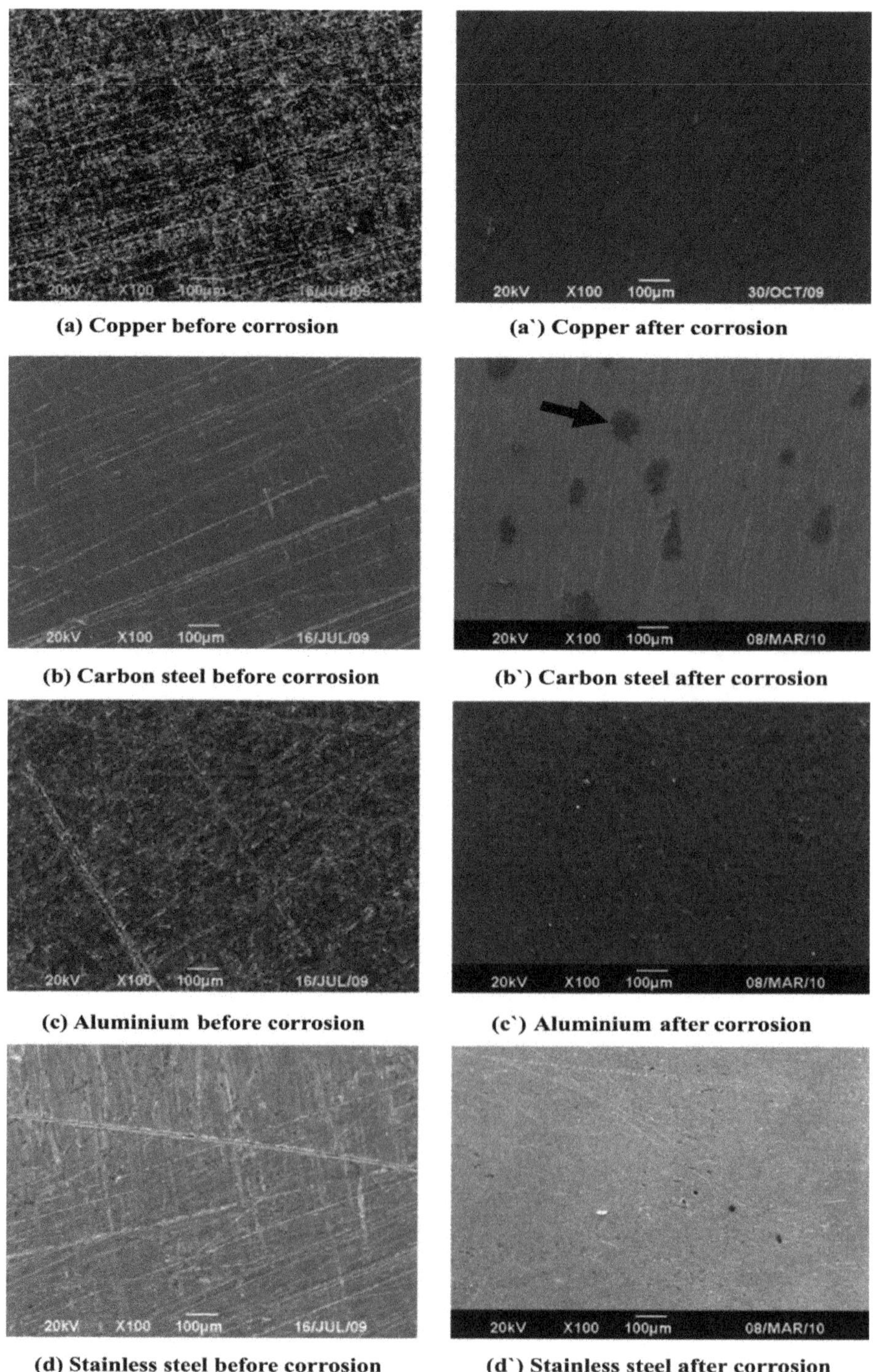

Figure 17. SEM image of the metal surface before and after immersion [71].

When the corrosion rates of AW 6060 aluminum alloy and E-Cu57 copper were assessed by ultralow sulfur diesel (ULSD) and rapeseed methyl ester (RME) for 600 h at 80 °C in the reaction of aluminum and biodiesel, $Al(OH)_3$ was found on the surface. Because of the presence of oxygen, another oxide layer Al_2O_3 was formed. In the case of copper, the compounds formed on the surface were $CuO/CuCO_3$ films followed by Cu_2O. The SEM images showed an increased concentration of biodiesel, which damaged the surface more and the pit generation can be seen to be increased, as in Figure 18 [72].

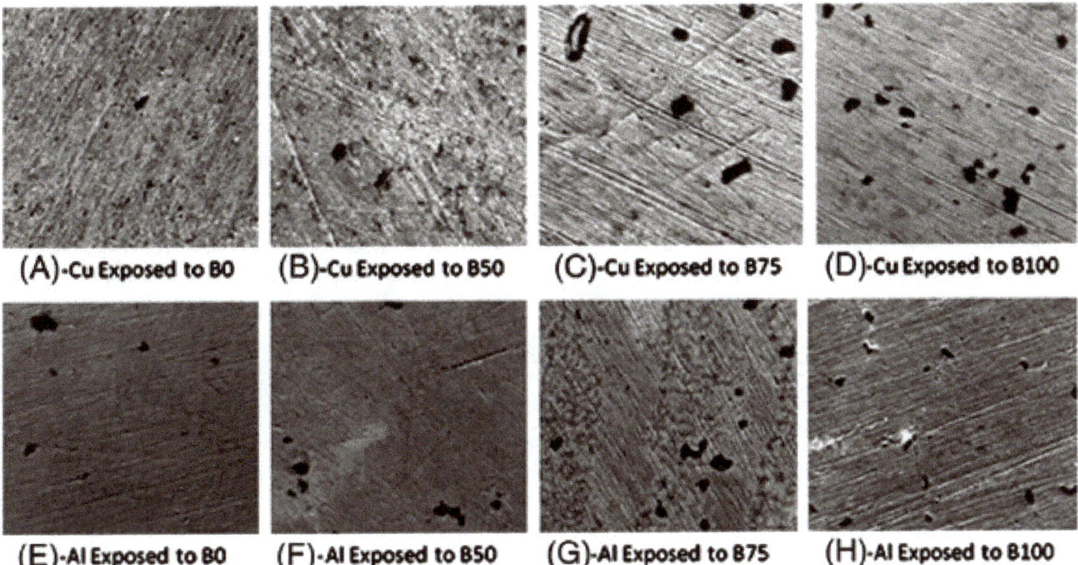

Figure 18. SEM photographs of surfaces exposed to fuel blends [72].

6.3. Characterization and Products of Corrosion Obtained by Sunflower Biodiesel

When CR of Al, Cu and MCS were assessed in sunflower biodiesel at RT and 60 °C for 3000 h, the XRD patterns of the materials confirmed the presence of base metal along with the little quantities of $FeCO_3$, $FeO(OH)$, Fe_2O_3, $Cu(OH)_2$, CuO and $AlO(OH)$. SEM image of aluminum showed that at RT no corrosion attack was found; however, at 60 °C, a protective layer covering the surface of the Al protected the metal surface from corrosion.

The SEM image of Cu in Figure 19 showed that at RT there was the visibility of a few small pit formations which indicated the initiation of corrosion, while at 60 °C, the formed pits were of larger size, which confirmed that the Cu had been corroded [74].

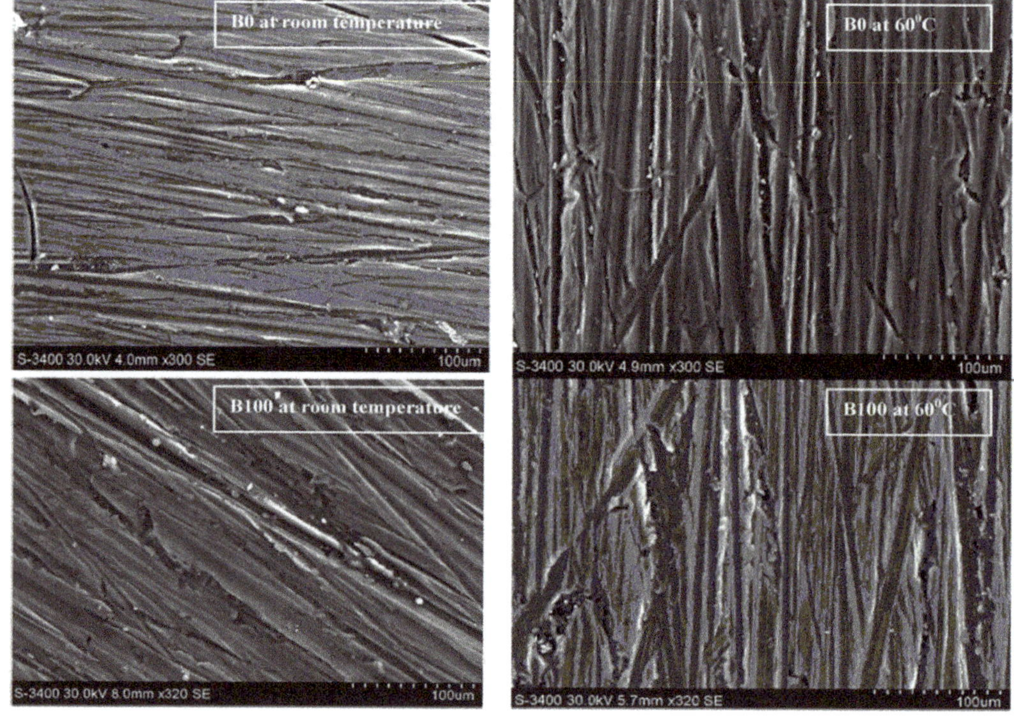

Figure 19. *Cont.*

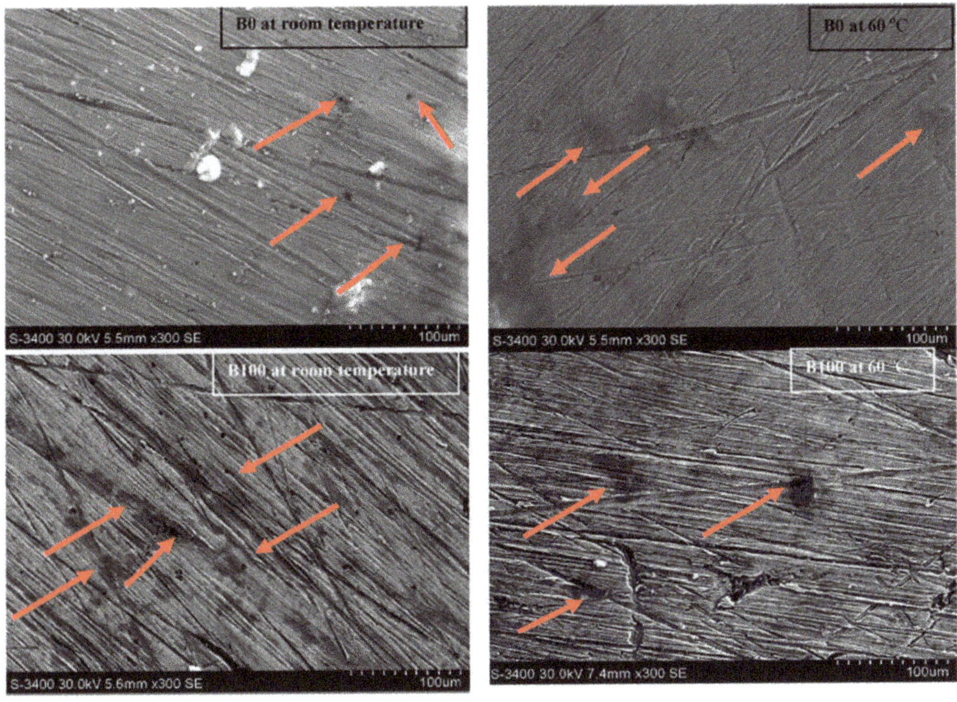

(iii)

Figure 19. SEM images of Al (**i**), Cu (**ii**) and MCS (**iii**) after immersions in diesel and biodiesel for 3000 h [74].

7. Application of Corrosion Inhibitors to Protect Corrosion

Jakeria et al. [69] used two organic corrosion inhibitors to assess the corrosion behaviors of MS and Cu when exposed to palm biodiesel. The inhibitors which were used include adenine and benzotriazole. Static immersion tests were performed at RT for 60 days. The CR of Cu had been reduced effectively by using benzotriazole than that of MS after 60 days. The CR of Cu dropped from 0.7495 to 0.0229 mg/cm^2. In the presence of adenine, the CR of Cu reduced from 0.7495 to 0.2512 mg/cm^2. There was little effect of adenine found in the reduction in the CR of MS. Therefore, it was concluded that benzotriazole provides better corrosion resistance as compared to adenine.

8. Future Recommendations

The biodiesel obtained from different sources has proved to be corrosive among the engine parts. Most of the studies to assess the corrosion rate of materials when immersed in biodiesel have been performed at higher blends of biodiesel i.e., at B50 or more. However, the usage of biodiesel along with diesel is not increased beyond B5 in most countries, and the reason behind this is the corrosiveness of biodiesel. Therefore, to enhance the usage of biodiesel in engines, the blends with percentages 10, 15 or 20 should be examined so that these can be further studied and implemented to be used in diesel engines along with diesel. The higher blends always showed that the biodiesel is corrosive and hence cannot be used in diesel engines. Therefore, some systematic studies are the requirement to implement a more percentage of biodiesel in blend with diesel.

As the compatibility of biodiesel with engine parts is not good, its use in automotive engines is limited. The engine parts are made from many materials involving steel and its alloys, aluminum, copper, brass and some other materials. Most of the studies have shown that copper and brass are the materials that have been most affected by biodiesel in terms of corrosion. Therefore, some alternate materials such as aluminum and stainless steel should be used to make parts that were previously made from copper, as stainless steel and aluminum have proved to be the least corrosive in biodiesel.

Additionally, to avoid corrosion, the materials can be coated with some useful coatings to prevent corrosion if it is necessary to use copper and brass parts of engines. These coatings will then reduce the intensity of corrosion of materials.

Corrosion inhibitors such as adenine and benzotriazole have shown good results in reducing the corrosion intensity of materials; however, these studies were performed for a limited period and hence they can be performed for longer durations and can be used in proper fractions to reduce the corrosion of materials in biodiesel.

9. Conclusions

Corrosion studies of materials, when coming in contact with biodiesel, are discussed in this study and it can be concluded that:

- Copper has is most affected materials in biodiesel in terms of corrosion followed by brass-, aluminum- and steel-based alloys, respectively. Additionally, the corrosion phenomena, surface morphology, mechanisms of corrosion and products of corrosion have been presented and it can be concluded that pitting is the most common type of corrosion that occurs from the use of biodiesel.
- Most of the materials produce their respective oxides in biodiesel and because of the presence of free oxygen.
- The immersion test method is a beneficial method for corrosion rate measurement.
- The main reasons of corrosion were the presence of unsaturated fatty acids, free water content, the biodiesel feedstock and condensation water on the surface of materials.
- With corrosion, the biodiesel degrades and hence results in an increased wear rate of the engine parts in contact with the biodiesel. Therefore, it is important to minimize corrosion affects produced by the use of biodiesel in engines.
- Most of the biodiesels have shown increased corrosion of materials when the concentration of biodiesel and duration of immersion or temperature was increased.

Author Contributions: Conceptualization, A.S. and A.A.; validation, M.M.Q., S.M.A.R. and M.H.H.; formal analysis, M.J.; investigation, H.M.A.J.; data curation, A.A.; writing—original draft preparation, A.S.; writing—review and editing, M.M.Q.; visualization, M.H.H.; supervision, A.A. All authors have read and agreed to the published version of the manuscript.

Funding: This research received no external funding.

Institutional Review Board Statement: Not applicable.

Informed Consent Statement: Not applicable.

Data Availability Statement: Not applicable.

Conflicts of Interest: The authors declare no conflict of interest.

Abbreviations

CR	Corrosion Rate		Spectroscopy	
SS	Stainless Steel		WSOB	Waste Sunflower Oil Biodiesel
Mg	Magnesium		FTIR	Fourier Transform Infrared
MS	Mild Steel		EDS	Energy Dispersive Spectroscopy
SEM	Scanning Electron Microscope		RT	Room Temperature
XRD	X-Ray Diffraction		HT	High Temperature
TAN	Total Acid Number		Cu	Copper
OM	Optical Microscope		BS	Brass
XPS	X-ray Photoelectron Spectroscopy		Al	Aluminum
Spectroscopy			CI	Cast Iron
AAS	Atomic Absorption Spectroscopy		MCS	Mild Carbon Steel

References

1. Aquino, I.P.; Hernandez, R.P.B.; Chicoma, D.L.; Pinto, H.P.F.; Aoki, I.V. Influence of light, temperature and metallic ions on biodiesel degradation and corrosiveness to copper and brass. *Fuel* **2012**, *102*, 795–807. [CrossRef]
2. McCormick, R.; Ratcliff, M.; Moens, L.; Lawrence, R. Several factors affecting the stability of biodiesel in standard accelerated tests. *Fuel Process. Technol.* **2007**, *88*, 651–657. [CrossRef]
3. Sadrolhosseini, A.R.; Moksin, M.M.; Yunus, W.M.M.; Talib, Z.A.; Abdi, M.M. Surface plasmon resonance detection of copper corrosion in biodiesel using polypyrrole-chitosan layer sensor. *Opt. Rev.* **2011**, *18*, 331–337. [CrossRef]
4. Ononiwu, I.; Adams, F.V.; Joseph, I.; Afolabi, A.S. Corrosion behaviour of mild steel in biodiesel prepared from ghee butter. In Proceedings of the World Congress on Engineering, London, UK, 1–3 July 2015.
5. Kralova, I.; Sjöblom, J. Biofuels–renewable energy sources: A review. *J. Dispers. Sci. Technol.* **2010**, *31*, 409–425. [CrossRef]
6. Kansedo, J.; Lee, K.T.; Bhatia, S. Biodiesel production from palm oil via heterogeneous transesterification. *Biomass Bioenergy* **2009**, *33*, 271–276. [CrossRef]
7. Antolın, G.; Tinaut, F.V.; Briceño, Y.; Castaño, V.; Pérez, C.; Ramírez, A.I. Optimisation of biodiesel production by sunflower oil transesterification. *Bioresour. Technol.* **2002**, *83*, 111–114. [CrossRef]
8. Qiu, F.; Li, Y.; Yang, D.; Li, X.; Sun, P. Biodiesel production from mixed soybean oil and rapeseed oil. *Appl. Energy* **2011**, *88*, 2050–2055. [CrossRef]
9. Issariyakul, T.; Kulkarni, M.G.; Meher, L.C.; Dalai, A.K.; Bakhshi, N.N. Biodiesel production from mixtures of canola oil and used cooking oil. *Chem. Eng. J.* **2008**, *140*, 77–85. [CrossRef]
10. Du, W.; Xu, Y.; Liu, D.; Zeng, J. Comparative study on lipase-catalyzed transformation of soybean oil for biodiesel production with different acyl acceptors. *J. Mol. Catal. B Enzym.* **2004**, *30*, 125–129. [CrossRef]
11. Veljković, V.B.; Biberdžić, M.O.; Banković-Ilić, I.B.; Djalović, I.G.; Tasić, M.B.; Nježić, Z.B.; Stamenković, O.S. Biodiesel production from corn oil: A review. *Renew. Sustain. Energy Rev.* **2018**, *91*, 531–548. [CrossRef]
12. Gürü, M.; Koca, A.; Özer, C.; Çınar, C.; Şahin, F. Biodiesel production from waste chicken fat based sources and evaluation with Mg based additive in a diesel engine. *Renew. Energy* **2010**, *35*, 637–643. [CrossRef]
13. Meher, L.; Sagar, D.V.; Naik, S. Technical aspects of biodiesel production by transesterification—A review. *Renew. Sustain. Energy Rev.* **2006**, *10*, 248–268. [CrossRef]
14. Sheehan, J.; Camobreco, V.; Duffield, J.; Graboski, M.; Shapouri, H. *An Overview of Biodiesel and Petroleum Diesel Life Cycles*; National Renewable Energy Lab. (NREL): Golden, CO, USA, 1998.
15. Knothe, G. Dependence of biodiesel fuel properties on the structure of fatty acid alkyl esters. *Fuel Process. Technol.* **2005**, *86*, 1059–1070. [CrossRef]
16. Espinosa, N.; Hösel, M.; Angmo, D.; Krebs, F.C. Solar cells with one-day energy payback for the factories of the future. *Energy Environ. Sci.* **2011**, *5*, 5117–5132. [CrossRef]
17. Kannan, N.; Vakeesan, D. Solar energy for future world: A review. *Renew. Sustain. Energy Rev.* **2016**, *62*, 1092–1105. [CrossRef]
18. Höök, M.; Tang, X. Depletion of fossil fuels and anthropogenic climate change—A review. *Energy Policy* **2013**, *52*, 797–809. [CrossRef]
19. Demirbaş, A. Bioenergy, Global Warming, and Environmental Impacts. *Energy Sources* **2004**, *26*, 225–236. [CrossRef]
20. McGlade, C.; Ekins, P. The geographical distribution of fossil fuels unused when limiting global warming to 2 °C. *Nat. Cell Biol.* **2015**, *517*, 187–190. [CrossRef] [PubMed]
21. Jaichandar, S.; Annamalai, K. The status of biodiesel as an alternative fuel for diesel engine—An Overview. *J. Sustain. Energy Environ.* **2011**, *2*, 71–75.
22. Klass, D.L. A critical assessment of renewable energy usage in the USA. *Energy Policy* **2003**, *31*, 353–367. [CrossRef]
23. Bowman, M.; Hilligoss, D.; Rasmussen, S.; Thomas, R. Biodiesel: A renewable and biodegradable fuel. *Hydrocarb. Process.* **2006**, *85*, 103.

24. Chen, K.-S.; Lin, Y.-C.; Hsieh, L.-T.; Lin, L.-F.; Wu, C.-C. Saving energy and reducing pollution by use of emulsified palm-biodiesel blends with bio-solution additive. *Energy* **2010**, *35*, 2043–2048. [CrossRef]

25. Clark, S.J.; Wagner, L.; Schrock, M.D.; Piennaar, P.G. Methyl and ethyl soybean esters as renewable fuels for diesel engines. *J. Am. Oil Chem. Soc.* **1984**, *61*, 1632–1638. [CrossRef]

26. Tian, Y.; Zhao, L.; Meng, H.; Sun, L.; Yan, J. Estimation of un-used land potential for biofuels development in (the) People's Republic of China. *Appl. Energy* **2009**, *86*, S77–S85. [CrossRef]

27. Yan, J.; Alvfors, P.; Eidensten, L.; Svedberg, G. A future for biomass. *Mech. Eng.* **1997**, *119*, 94.

28. Ghobadian, B.; Rahimi, H.; Nikbakht, A.; Najafi, G.; Yusaf, T. Diesel engine performance and exhaust emission analysis using waste cooking biodiesel fuel with an artificial neural network. *Renew. Energy* **2009**, *34*, 976–982. [CrossRef]

29. Ramírez-Verduzco, L.F.; Rodríguez-Rodríguez, J.E.; Jaramillo-Jacob, A.D.R. Predicting cetane number, kinematic viscosity, density and higher heating value of biodiesel from its fatty acid methyl ester composition. *Fuel* **2012**, *91*, 102–111. [CrossRef]

30. Gülüm, M.; Bilgin, A. Density, flash point and heating value variations of corn oil biodiesel–diesel fuel blends. *Fuel Process. Technol.* **2015**, *134*, 456–464. [CrossRef]

31. An, H.; Yang, W.; Chou, S.; Chua, K. Combustion and emissions characteristics of diesel engine fueled by biodiesel at partial load conditions. *Appl. Energy* **2012**, *99*, 363–371. [CrossRef]

32. Muñoz, M.; Moreno, F.; Monné, C.; Morea, J.; Terradillos, J. Biodiesel improves lubricity of new low sulphur diesel fuels. *Renew. Energy* **2011**, *36*, 2918–2924. [CrossRef]

33. Fazal, M.; Haseeb, A.; Masjuki, H. Comparative corrosive characteristics of petroleum diesel and palm biodiesel for automotive materials. *Fuel Process. Technol.* **2010**, *91*, 1308–1315. [CrossRef]

34. Popovicheva, O.; Engling, G.; Lin, K.-T.; Persiantseva, N.; Timofeev, M.; Kireeva, E.D.; Völk, P.; Hubert, A.; Wachtmeister, G. Diesel/biofuel exhaust particles from modern internal combustion engines: Microstructure, composition, and hygroscopicity. *Fuel* **2015**, *157*, 232–239. [CrossRef]

35. Ghazali, W.N.M.W.; Mamat, R.; Masjuki, H.; Najafi, G. Effects of biodiesel from different feedstocks on engine performance and emissions: A review. *Renew. Sustain. Energy Rev.* **2015**, *51*, 585–602. [CrossRef]

36. Rakopoulos, C.; Rakopoulos, D.; Hountalas, D.; Giakoumis, E.; Andritsakis, E. Performance and emissions of bus engine using blends of diesel fuel with bio-diesel of sunflower or cottonseed oils derived from Greek feedstock. *Fuel* **2008**, *87*, 147–157. [CrossRef]

37. Haseeb, A.; Sia, S.; Fazal, M.; Masjuki, H. Effect of temperature on tribological properties of palm biodiesel. *Energy* **2010**, *35*, 1460–1464. [CrossRef]

38. Rashid, U.; Anwar, F.; Moser, B.R.; Knothe, G. Moringa oleifera oil: A possible source of biodiesel. *Bioresour. Technol.* **2008**, *99*, 8175–8179. [CrossRef]

39. Demirbas, A. Progress and recent trends in biofuels. *Prog. Energy Combust. Sci.* **2007**, *33*, 1–18. [CrossRef]

40. Goldemberg, J.; Guardabassi, P. Are biofuels a feasible option? *Energy Policy* **2009**, *37*, 10–14. [CrossRef]

41. Dharma, S.; Silitonga, A.; Shamsuddin, A.; Sebayang, A.H.; Milano, J.; Sebayang, R.; Sarjianto; Ibrahim, H.; Bahri, N.; Ginting, B.; et al. Properties and corrosion behaviors of mild steel in biodiesel-diesel blends. *Energy Sources Part A Recover. Util. Environ. Eff.* **2019**, 1–13. [CrossRef]

42. Thangavelu, S.K.; Chelladorai, P.; Ani, F.N. Corrosion Behaviour of Carbon Steel in Biodiesel–Diesel–Ethanol (BDE) Fuel Blend. In Proceedings of the MATEC Web of Conferences, 2015 4th International Conference on Engineering and Innovative Materials (ICEIM 2015) EDP Sciences, Penang, Malaysia, 20 October 2015.

43. Canakci, M.; Ozsezen, A.N.; Arcaklioglu, E.; Erdil, A. Prediction of performance and exhaust emissions of a diesel engine fueled with biodiesel produced from waste frying palm oil. *Expert Syst. Appl.* **2009**, *36*, 9268–9280. [CrossRef]

44. Canakci, M.; Van Gerpen, J.H. Comparison of engine performance and emissions for petroleum diesel fuel, yellow grease biodiesel, and soybean oil biodiesel. *Trans. ASAE* **2003**, *46*, 937. [CrossRef]

45. Labeckas, G.; Slavinskas, S. The effect of rapeseed oil methyl ester on direct injection Diesel engine performance and exhaust emissions. *Energy Convers. Manag.* **2006**, *47*, 1954–1967. [CrossRef]

46. Kalligeros, S.; Zannikos, F.; Stournas, S.; Lois, E.; Anastopoulos, G.; Teas, C.; Sakellaropoulos, F. An investigation of using biodiesel/marine diesel blends on the performance of a stationary diesel engine. *Biomass Bioenergy* **2003**, *24*, 141–149. [CrossRef]

47. Dorado, M. Exhaust emissions from a Diesel engine fueled with transesterified waste olive oil⋆. *Fuel* **2003**, *82*, 1311–1315. [CrossRef]

48. Valente, O.S.; Da Silva, M.J.; Pasa, V.M.D.; Belchior, C.R.P.; Sodré, J.R. Fuel consumption and emissions from a diesel power generator fuelled with castor oil and soybean biodiesel. *Fuel* **2010**, *89*, 3637–3642. [CrossRef]

49. Pramanik, K. Properties and use of jatropha curcas oil and diesel fuel blends in compression ignition engine. *Renew. Energy* **2003**, *28*, 239–248. [CrossRef]

50. Sureshkumar, K.; Velraj, R.; Ganesan, R. Performance and exhaust emission characteristics of a CI engine fueled with Pongamia pinnata methyl ester (PPME) and its blends with diesel. *Renew. Energy* **2008**, *33*, 2294–2302. [CrossRef]

51. Nabi, M.N.; Hoque, S.N. Biodiesel production from linseed oil and performance study of a diesel engine with diesel bio-diesel. *J. Mech. Eng.* **2008**, *39*, 40–44. [CrossRef]

52. Holser, R.A.; Harry-O'Kuru, R. Transesterified milkweed (Asclepias) seed oil as a biodiesel fuel. *Fuel* **2006**, *85*, 2106–2110. [CrossRef]

53. Haseeb, A.; Fazal, M.; Jahirul, M.; Masjuki, H. Compatibility of automotive materials in biodiesel: A review. *Fuel* **2011**, *90*, 922–931. [CrossRef]

54. Van Wechem, G.; Beunk, G.; Van Den Elst, F.; Van Der Ploeg, A. Combustion engine with fuel injection system, and a spray valve for such an engine. U.S. Patent 4898142A, 6 February 1990.

55. Haseeb, A.S.M.A.; Masjuki, H.H.; Ann, L.J.; Fazal, M.A. Corrosion characteristics of copper and leaded bronze in palm biodiesel. *Fuel Process. Technol.* **2010**, *91*, 329–334. [CrossRef]

56. Arslan, R. Emission characteristics of a diesel engine using waste cooking oil as biodiesel fuel. *Afr. J. Biotechnol.* **2011**, *10*, 3790–3794.

57. Cheng, S.; Lloyd, I.K.; Kahn, M. Modification of Surface Texture by Grinding and Polishing Lead Zirconate Titanate Ceramics. *J. Am. Ceram. Soc.* **1992**, *75*, 2293–2296. [CrossRef]

58. Kagwade, S.V.; Clayton, C.R.; Chidambaram, D.; Du, M.L.; Chiang, F.P. The Influence of Acetone Degreasing on the Corrosion Behavior of AA2024-T3. *J. Electrochem. Soc.* **2000**, *147*, 4125–4130. [CrossRef]

59. Internasional, A. ASTM G31-72: Standard Practice for Laboratory Immersion Corrosion Testing of Metals. United State. 2004. Available online: https://www.astm.org/DATABASE.CART/HISTORICAL/G31-72R04.htm (accessed on 3 March 2021).

60. Ahn, S.; Choi, Y.; Kim, J.; Han, J. A study on corrosion resistance characteristics of PVD Cr-N coated steels by electrochemical method. *Surf. Coat. Technol.* **2002**, *150*, 319–326. [CrossRef]

61. Cabeza, L.F.; Roca, J.; Nogués, M.; Mehling, H.; Hiebler, S. Immersion corrosion tests on metal-salt hydrate pairs used for latent heat storage in the 48 to 58 °C temperature range. *Mater. Corros.* **2002**, *53*, 902–907. [CrossRef]

62. Akhabue, C.; Aisien, F.; Ojo, C. The effect of Jatropha oil biodiesel on the corrosion rates of aluminium and mild carbon steel. *Biofuels* **2014**, *5*, 545–550. [CrossRef]

63. Sterpu, A.-E.; Dumitru, A.I.; Popa, M.-F. Corrosion behavior of steel in biodiesel of different origin. *An. Univ. Ovidius Constanta Ser. Chim.* **2012**, *23*, 143–148. [CrossRef]

64. Fazal, M.; Haseeb, A.; Masjuki, H. Degradation of automotive materials in palm biodiesel. *Energy* **2012**, *40*, 76–83. [CrossRef]

65. Thangavelu, S.K.; Piraiarasi, C.; Ahmed, A.; Ani, F. Corrosion Behavior of Copper in Biodiesel-Diesel-Bioethanol (BDE). *Adv. Mater. Res.* **2015**, *1098*, 44–50. [CrossRef]

66. Jin, D.; Zhou, X.; Wu, P.; Jiang, L.; Ge, H. Corrosion behavior of ASTM 1045 mild steel in palm biodiesel. *Renew. Energy* **2015**, *81*, 457–463. [CrossRef]

67. Fazal, M.; Haseeb, A.; Masjuki, H. Effect of temperature on the corrosion behavior of mild steel upon exposure to palm biodiesel. *Energy* **2011**, *36*, 3328–3334. [CrossRef]

68. Chew, K.; Haseeb, A.; Masjuki, H.; Fazal, M.; Gupta, M. Corrosion of magnesium and aluminum in palm biodiesel: A comparative evaluation. *Energy* **2013**, *57*, 478–483. [CrossRef]

69. Jakeria, M.R.; Fazal, M.A.; Haseeb, A.S.M.A. Effect of corrosion inhibitors on corrosiveness of palm biodiesel. *Corros. Eng. Sci. Technol.* **2015**, *50*, 56–62. [CrossRef]

70. Kaul, S.; Saxena, R.; Kumar, A.; Negi, M.; Bhatnagar, A.; Goyal, H.; Gupta, A. Corrosion behavior of biodiesel from seed oils of Indian origin on diesel engine parts. *Fuel Process. Technol.* **2007**, *88*, 303–307. [CrossRef]

71. Hu, E.; Xu, Y.; Hu, X.; Pan, L.; Jiang, S. Corrosion behaviors of metals in biodiesel from rapeseed oil and methanol. *Renew. Energy* **2012**, *37*, 371–378. [CrossRef]

72. Norouzi, S.; Eslami, F.; Wyszynski, M.L.; Tsolakis, A. Corrosion effects of RME in blends with ULSD on aluminium and copper. *Fuel Process. Technol.* **2012**, *104*, 204–210. [CrossRef]

73. Cursaru, D.; Mihai, S. Corrosion Behaviour of Automotive Materials in Biodiesel from Sunflower Oil. *Rev. Chim.* **2012**, *63*, 149–158.

74. Samuel, O.D.; Gulum, M. Mechanical and corrosion properties of brass exposed to waste sunflower oil biodiesel-diesel fuel blends. *Chem. Eng. Commun.* **2019**, *206*, 682–694. [CrossRef]

75. Cursaru, D.-L.; Brănoiu, G.; Ramadan, I.; Miculescu, F. Degradation of automotive materials upon exposure to sunflower biodiesel. *Ind. Crop. Prod.* **2014**, *54*, 149–158. [CrossRef]

76. Díaz-Ballote, L.; López-Sansores, J.; Maldonado-López, L.; Garfias-Mesias, L. Corrosion behavior of aluminum exposed to a biodiesel. *Electrochem. Commun.* **2009**, *11*, 41–44. [CrossRef]

77. Lu, Q.; Zhang, J.; Zhu, X. Corrosion properties of bio-oil and its emulsions with diesel. *Chin. Sci. Bull.* **2008**, *53*, 3726–3734. [CrossRef]

78. Román, A.S.; Méndez, C.M.; Ares, A.E. Corrosion Resistance of Stainless Steels in Biodiesel. In *Shape Casting: 6th International Symposium*; Springer: Cham, Switzerland, 2016.

79. Geller, D.P.; Adams, T.T.; Goodrum, J.W.; Pendergrass, J. Storage stability of poultry fat and diesel fuel mixtures: Specific gravity and viscosity. *Fuel* **2008**, *87*, 92–102. [CrossRef]

80. Aghbashlo, M.; Tabatabaei, M.; Khalife, E.; Najafi, B.; Mirsalim, S.M.; Gharehghani, A.; Mohammadi, P.; Dadak, A.; Shojaei, T.R.; Khounani, Z. A novel emulsion fuel containing aqueous nano cerium oxide additive in diesel–biodiesel blends to improve diesel engines performance and reduce exhaust emissions: Part II—Exergetic analysis. *Fuel* **2017**, *205*, 262–271. [CrossRef]

81. Aghbashlo, M.; Tabatabaei, M.; Mohammadi, P.; Mirzajanzadeh, M.; Ardjmand, M.; Rashidi, A. Effect of an emission-reducing soluble hybrid nanocatalyst in diesel/biodiesel blends on exergetic performance of a DI diesel engine. *Renew. Energy* **2016**, *93*, 353–368. [CrossRef]

82. Aghbashlo, M.; Tabatabaei, M.; Mohammadi, P.; Pourvosoughi, N.; Nikbakht, A.M.; Goli, S.A.H. Improving exergetic and sustainability parameters of a DI diesel engine using polymer waste dissolved in biodiesel as a novel diesel additive. *Energy Convers. Manag.* **2015**, *105*, 328–337. [CrossRef]

83. Aghbashlo, M.; Tabatabaei, M.; Rastegari, H.; Ghaziaskar, H.S. Exergy-based sustainability analysis of acetins synthesis through continuous esterification of glycerol in acetic acid using Amberlyst®36 as catalyst. *J. Clean. Prod.* **2018**, *183*, 1265–1275. [CrossRef]

84. Ahmmad, M.S.; Haji Hassan, M.B.; Kalam, M.A. Comparative corrosion characteristics of automotive materials in Jatropha biodiesel. *Int. J. Green Energy* **2018**, *15*, 393–399. [CrossRef]

85. Aksoy, F. Alkaline catalyzed biodiesel production from safflower (Carthamus tinctorius L.) oil: Optimization of parameters and determination of fuel properties. *Energy Sources Part A Recover. Util. Environ. Eff.* **2016**, *38*, 835–841. [CrossRef]

86. Al-Dawody, M.F.; Bhatti, S. Optimization strategies to reduce the biodiesel NOx effect in diesel engine with experimental verification. *Energy Convers. Manag.* **2013**, *68*, 96–104. [CrossRef]

87. Amaya, A.; Piamba, O.; Olaya, J. Improvement of Corrosion Resistance for Gray Cast Iron in Palm Biodiesel Application Using Thermoreactive Diffusion Niobium Carbide (NbC) Coating. *Coatings* **2018**, *8*, 216. [CrossRef]

88. Amgain, K.; Subedi, B.N.; Joshi, S.; Bhattarai, J. Investigation on the effect of tinospora cordifolia plant extract as a green corrosion inhibitor to aluminum and copper in biodiesel and its blend. In Proceedings of the NIGIS* CORCON 2018, Jaipur, India, 30 September–3 October 2018. Paper No. PP19.

89. Ashraful, A.; Masjuki, H.; Kalam, M.; Rashedul, H.; Sajjad, H.; Abedin, M. Influence of anti-corrosion additive on the performance, emission and engine component wear characteristics of an IDI diesel engine fueled with palm biodiesel. *Energy Convers. Manag.* **2014**, *87*, 48–57. [CrossRef]

Article

Effect of Additivized Biodiesel Blends on Diesel Engine Performance, Emission, Tribological Characteristics, and Lubricant Tribology

M. A. Mujtaba [1,2,*], H. H. Masjuki [1,3], M. A. Kalam [1,*], Fahad Noor [2], Muhammad Farooq [2], Hwai Chyuan Ong [4], M. Gul [1,5], Manzoore Elahi M. Soudagar [1], Shahid Bashir [6], I. M. Rizwanul Fattah [4,*] and L. Razzaq [2]

1   Department of Mechanical Engineering, Center for Energy Science, University of Malaya, Kuala Lumpur 50603, Malaysia; masjuki@um.edu.my (H.H.M.); mustabshirha@bzu.edu.pk (M.G.); manzoor@siswa.um.edu.my (M.E.M.S.)
2   Department of Mechanical, Mechatronics and Manufacturing Engineering (New Campus), University of Engineering and Technology Lahore, Lahore 54000, Pakistan; f.noor@uet.edu.pk (F.N.); engr.farooq@uet.edu.pk (M.F.); luqmanrazzaq@uet.edu.pk (L.R.)
3   Department of Mechanical Engineering, Faculty of Engineering, The International Islamic University Malaysia, Kuala Lumpur 50728, Malaysia
4   School of Information, Systems and Modelling, Faculty of Engineering and IT, University of Technology Sydney, Ultimo, New South Wales 2007, Australia; HwaiChyuan.Ong@uts.edu.au
5   Department of Mechanical Engineering, Faculty of Engineering and Technology, Bahauddin Zakariya University, Multan 60000, Pakistan
6   Department of Physics, Center of Ionics, Faculty of Science University of Malaya Malaysia, University of Malaya, Kuala Lumpur 50603, Malaysia; shahidbashirbaig@um.edu.my
*   Correspondence: m.mujtaba@uet.edu.pk (M.A.M.); kalam@um.edu.my (M.A.K.); rizwanul.buet@gmail.com or IslamMdRizwanul.Fattah@uts.edu.au (I.M.R.F.)

Received: 5 June 2020; Accepted: 26 June 2020; Published: 1 July 2020

Abstract: This research work focuses on investigating the lubricity and analyzing the engine characteristics of diesel–biodiesel blends with fuel additives (titanium dioxide (TiO_2) and dimethyl carbonate (DMC)) and their effect on the tribological properties of a mineral lubricant. A blend of palm–sesame oil was used to produce biodiesel using ultrasound-assisted transesterification. B30 (30% biodiesel + 70% diesel) fuel was selected as the base fuel. The additives used in the current study to prepare ternary fuel blends were TiO_2 and DMC. B30 + TiO_2 showed a significant reduction of 6.72% in the coefficient of friction (COF) compared to B30. B10 (Malaysian commercial diesel) exhibited very poor lubricity and COF among all tested fuels. Both ternary fuel blends showed a promising reduction in wear rate. All contaminated lubricant samples showed an increment in COF due to the dilution of combustible fuels. Lub + B10 (lubricant + B10) showed the highest increment of 42.29% in COF among all contaminated lubricant samples. B30 + TiO_2 showed the maximum reduction (6.76%) in brake-specific fuel consumption (BSFC). B30 + DMC showed the maximum increment (8.01%) in brake thermal efficiency (BTE). B30 + DMC exhibited a considerable decline of 32.09% and 25.4% in CO and HC emissions, respectively. The B30 + TiO_2 fuel blend showed better lubricity and a significant improvement in engine characteristics.

Keywords: high frequeny reciprocating rig; palm-sesame biodiesel; nanoparticle additives; four-ball tribo tester; engine characteristics; tribological characteristics

---

1. Introduction

Global energy demand is gradually increasing owing to a significant increase in population growth. The transport sector is one of the major consumers of energy, which is the backbone of every country. Around half of the petroleum products are used to fulfil the high energy demand in the transport sector. However, as the fossil fuel reserves deplete gradually, alternative fuels are expected to meet energy demands in future [1]. Among alternative fuels, biodiesel shows a significant reduction in some of the critical exhaust emissions (CO and HC) due to its superior properties such as cetane number, oxygen content, flash point, etc. to those of petroleum-based diesel fuel [2–4]. However, pure biodiesel suffers from poor brake thermal efficiency (BTE) due to lower calorific values which can be alleviated by blending biodiesel with diesel [5,6]. Biodiesel commercialization has some limitations due to poor cold flow properties, higher NOx emissions, and poor oxidation stability, which can be resolved by mixing different additives such as nanoparticles and oxygenated compounds [7–9]. Malaysia is expected to adopt B30 (30% palm biodiesel in diesel) by 2025 [10]. Indonesia already introduced the B30 biodiesel as of 2020 [11]. At present, Indonesia and Malaysia are exporting palm oil to Asia and Europe for the production of biodiesel to meet energy demands which are expected to increase in future [12]. However, palm oil has very poor cold flow properties because of saturated fatty acids deteriorating its use in cold climate countries [13]. Mujtaba et al. [14] reported that sesame oil (SO) is the best-suited vegetable oil among all other feedstocks for the improvement of cold flow properties, as well as oxidation stability. SO exhibits very good cold flow properties due to high unsaturation. The oxidation stability of SO is higher despite high unsaturation due to antioxidants that occur naturally in SO [15].

The lubricity of fuel is a very important parameter that should be accounted for when selecting fuel for engine applications. Most diesel engine components are self-lubricated with fuel such as the fuel pump, fuel injector, etc. Petroleum diesel has very low lubricity due to the elimination of polar compounds during the desulfurization process [16]. These compounds assist fuel by making a lubricating protective layer between metallic contacts to minimize wear and friction. Lubrication is also very crucial to enhance the overall effectiveness of engine parts [17,18]. The addition of biodiesel in petroleum diesel improves the lubricity of diesel–biodiesel blends which results in lower wear scar diameter (WSD) and coefficient of friction (COF). Few researchers investigated the effect of biodiesel on the lubricity of diesel–biodiesel blends [19–21]. Many researchers used nanoparticles and oxygenated additives with methyl ester and diesel fuel blends to enhance the diesel engine characteristics. Very few studies were carried out on the lubricity of these ternary fuel blends used for improving engine characteristics.

A lubricant film reduces friction and wear; consequently, efficiency increases. According to previous literature, the lubricant is contaminated with fuel up to 5% due to crankcase dilution [20,22]. After dilution, lubricant properties are altered, which directly affects the tribological properties. Few researchers investigated the effect of biodiesel dilution with lubricant on its tribological characteristics [23–25]. Arumugam et al. [26] reported that rapeseed-bio lubricant contaminated with 10% rapeseed biodiesel (B20) fuel showed less wear and friction than commercial synthetic lubricant contaminated with 10% diesel fuel tested using a pin-on-disc apparatus with engine cylinder liner–piston ring combinations. Dhar et al. [19] investigated the tribology of lubricating oil of a diesel engine run on a Karanja biodiesel (B20) blend and mineral-based diesel during an endurance test of 200 h. Their investigation proved that lubricating oil obtained from the biodiesel-fueled engine contained a significantly high amount of wear, trace metals, resinous, ash content, and soot than the lubricating oil from the mineral diesel-fueled engine. Hence, 20% Karanja biodiesel (B20) caused more deterioration of lubricating oil than mineral diesel. Singh et al. [25] also found that more than 5–8% contamination of 100% pure moringa biodiesel fuel in lubricant enhances the lubricity but more than 8% contamination of moringa biodiesel fuel increased the wear rate considerably. Similarly, Maleque et al. [23] and Sulek et al. [24] reported that 5% of biodiesel fuel dilution in commercial lubricant decreased wear rate. A study should be conducted on the contamination of lubricant due to crank dilution with combustible ternary fuel blends. Many researchers did not pay any attention to lubricant degradation due to dilution with diesel–biodiesel–fuel additives. A tribological study of ternary fuel blends should be conducted before engine application to ensure its effect on degradation of the lubricant.

In this investigation, a 50:50 ratio of palm and SO was used to prepare the biodiesel using ultrasound-assisted transesterification to improve the physicochemical properties (oxidation stability and cold flow properties) of biodiesel. The main objectives of this work were (1) to investigate the lubricity of fuel additives used for enhancement in overall engine characteristics, (2) to investigate the effect of ternary fuel blend contamination with mineral lubricant on its tribological properties, and (3) to investigate the effect of ternary fuel blends on engine performance and emission characteristics.

2. Materials and Methods

Palm oil was procured from Malaysia and SO was obtained from Pakistan. In the current study, a mixture of palm–SO was used to prepare the biodiesel. According to ASTM D6079-11 dimensions, AISI 52100 Chrome hard polished steel balls with a diameter of 6.2 mm, 15-mm SAE-AMS 6440 steel smooth diamond polish discs, and 12.7-mm-diameter AISI 52100 steel balls having hardness 64–66 Rc were procured from the local market.

2.1. Biodiesel Production

A 50:50 ratio of palm–SO was used to prepare the biodiesel using Q500 Sonicator (QSONICA, Newtown, CT, USA) ultrasound equipment as shown in Figure 1. Ultrasound-assisted transesterification was performed under the following operating conditions: time (38.96 min), duty cycle (59.52%), temperature (60 °C), CH_3OH to palm–SO M ratio of 60 vol./vol.%, and potassium hydroxide used as a catalyst with a quantity of 0.70 wt.% [27]. The physicochemical properties of P50S50 biodiesel are presented in Table 1. The physicochemical properties of palm–sesame blend (50:50) methyl esters were estimated in accordance with biodiesel standard methods ASTM D6751 and EN 14214.

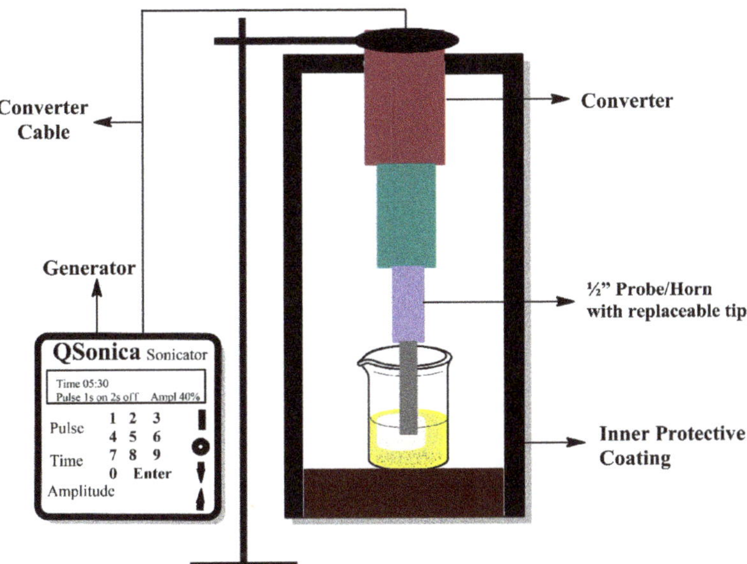

Figure 1. Ultrasound equipment for biodiesel production and sonication of P50S50 blended oil with nanoparticles.

Table 1. Physicochemical properties of P50S50 biodiesel.

Properties of Test Fuel	P50S50	Malaysian Diesel	Equipment	Accuracy
Kinematic viscosity at 40 °C (mm^2/s)	4.43	2.88	SVM 3000, (Anton Paar, Graz, Austria)	±0.35% mm^2/s
Acid Value (mg KOH/g)	0.375	0.152	-	-
Density at 15 °C (kg/m^3)	881	838.96	SVM 3000, (Anton Paar, Graz, Austria)	±0.1 kg/m^3
High heating value (MJ/kg)	41.24	45.67	C2000 basic calorimeter, IKA, Staufen, Germany	±0.1% MJ/kg
Cetane number	53.37	48.50	-	-
Cold filter plugging point (°C)	−1.714	0	CFFP NTL 450, Compass Instruments, IL, USA	-
Flash point (°C)	>151	77.90	Pensky-Martens closed cup tester NPM 440, Normalab, Valliquerville, France	±0.1 °C
Cloud point (°C)	7.82	2.05	Cloud and pour point tester NTE 450, Normalab, Valliquerville, France	±0.1 °C
Pour point (°C)	3.821	2.1	Cloud and pour point tester NTE 450, Normalab, Valliquerville, France	-
Oxidation stability (h)	6.89	13.20	873 Biodiesel Rancimat, Metrohm, Herisau, Switzerland	±0.01 h

2.2. Fuel Samples Preparation

2.2.1. Fuel Sample Preparation for HFRR and Engine Test Rig

Various fuel samples were prepared to study the lubricity of fuel and the effect of ternary fuel blends on diesel engine characteristics. The prepared fuel samples were compared with commercially available Malaysian diesel (B10). Pure B100 biodiesel was produced using the ultrasound technique. Then, 20% P50S50 biodiesel was mixed with Malaysian commercial diesel to prepare the B30 fuel blend. Additionally, two ternary fuel additive blends were prepared to study the tribological behavior. B30 fuel was mixed with 20% dimethyl carbonate (DMC) (by volume) and stirred at a standard speed 2000 rpm for half an hour to achieve a homogeneous blend. Ultrasound Sonicator Q500 Sonicator (QSONICA, Newtown, USA) as shown in Figure 1 was used to prepare the nanoparticle (TiO_2) and B30 ternary fuel blend. The B30 with 100 ppm TiO_2 nanoparticle fuel sample was prepared at a stirring speed of 900 rpm for 30 min on a magnetic stirrer and sonicated to disperse the nanoparticles at a frequency of 20 Hz for 20 min at an amplitude of 30%. The physicochemical properties were measured and are shown in Table 2.

Table 2. Properties of tested fuel samples. DMC—dimethyl carbonate.

Test Fuel Blends	Density at 15 °C	Kinematic Viscosity at 40 °C	Calorific Value	Viscosity Index
	kg/m^3	mm^2/s	MJ/kg	
B10	855.9	3.153	43.92	81.2
B100 (P50S50 biodiesel)	880	4.420	41.25	186.2
B30	852.6	3.348	43.14	164.2
B30 + DMC	878	2.457	39.78	-
B30 + TiO_2	853	3.364	42.93	208.3

2.2.2. Lubricant Sample Preparation

Firstly, 5% of each of the above-mentioned fuels in Table 2 was mixed with commercial lubricant SAE-40 using a magnetic stirrer at 900 rpm speed for 30 min, as 5% lubricant mixing with fuel happened due to dilution of the crankcase. The physicochemical properties of reference lubricant SAE-40 and all other lubricant samples with different fuels were measured using the viscometer (SVM 3000) as shown in Table 3.

Table 3. Physicochemical properties of tested lubricant fuel samples.

Lubricant Samples	Physicochemical Properties of Lubricant Samples		
	Density at 15 °C (kg/m³)	Viscosity at 40 °C (mm²/s)	Viscosity Index
100% Lubricant (SAE 40)	873.7	87,022	201.3
Lubricant + 5% B10	872.0	71,072	205.3
Lubricant + 5% B30	873.3	76,652	221.6
Lubricant + 5% B30 + DMC	872.0	64,509	-
Lubricant + 5% B30 + TiO₂	872.1	69,311	205.4

2.3. Experimental Set-Up

2.3.1. Diesel Engine Set-Up

The Yanmar (TF 120M) single-cylinder and radiator cooled diesel engine as shown in Figure 2 was used to study the effect of ternary fuel blends on diesel engine performance and emission characteristics. The diesel engine parameters were as follows: maximum power (7.7 kW), compression ratio (17.7), maximum rpm (2400 rpm), injection timing (17° before top dead center (BTDC)), and injection pressure (200 kg/cm²). The emissions of the engine exhaust gases such as CO, HC, and NOx were measured utilizing the BOSCH BEA 350 gas analyzer. All experiments were done in triplicate. Error bars are presented along with the data points in the figures. The accuracies of the measurements for diesel engine characteristics are mentioned in Table 4 The overall uncertainty of diesel engine experiments was calculated using Equation (1).

$$
\begin{aligned}
\text{Overall uncertainity} \ &= \ \sqrt{\text{Uncertainity \% of } \left(BSFC^2 + BTE^2 + BP^2 + CO^2 + NOx^2 + HC^2 \right)} \\
&= \ \sqrt{\left((1.31)^2 + (0.324)^2 + (0.667)^2 + (1)^2 + (1.3)^2 + (1)^2 \right)} \\
&= \ \pm\, 2.44\%.
\end{aligned}
\tag{1}
$$

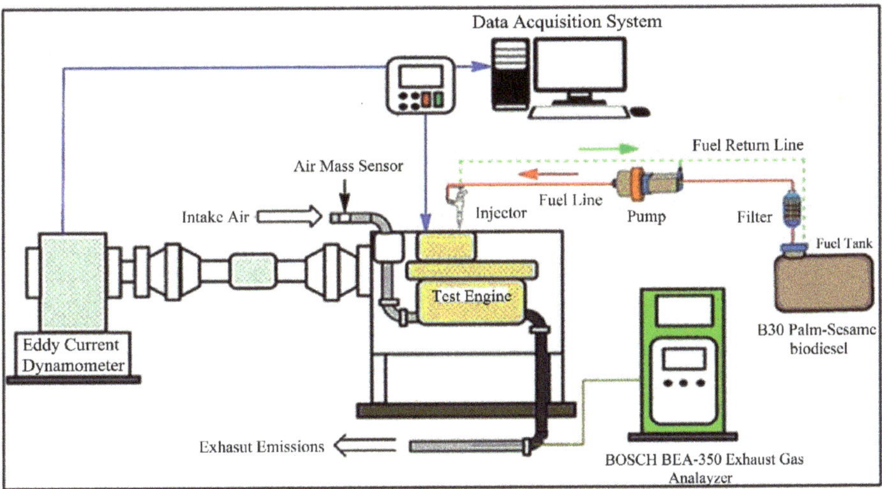

Figure 2. Schematic view of diesel engine set-up.

Table 4. Accuracies of measurements used for experiments.

Measurement	Measurement Range	Accuracy (±)
Speed	60–10,000 rpm	±10 rpm
Load	±120 Nm	±0.1 Nm
Flow Measurement	0.5–36 L/h	±0.01 L/h
CO	0–10 vol.%	±0.001 vol.%
HC	0–9999 ppm	±1 ppm
NOx	0–5000 ppm	±1 ppm

2.3.2. HFRR Test Rig

HFRR equipment from DUCOM (Model: TR-281-M8) as shown in Figure 3 was used to study the lubricity of tested fuel samples. The testing specimen plates were prepared by cutting 15 mm × 15 mm pieces. The specimens were polished with silicon carbide papers using the polishing machine. A ball on a test specimen plate was used to analyze the tribological behavior of fuel samples. Steel ball slides on the steel specimen plate were submerged in 5.0 ± 0.2 mL of fuel sample in a reciprocating motion with 2.0 ± 0.02 mm stroke length at a frequency 10.0 ± 1 Hz for 70 min with an applied load of 5 ± 0.01 N. Fuel temperature was constant at 60 ± 2 °C during the tribology test. All operating conditions followed the standard test method ASTM D6079-11.

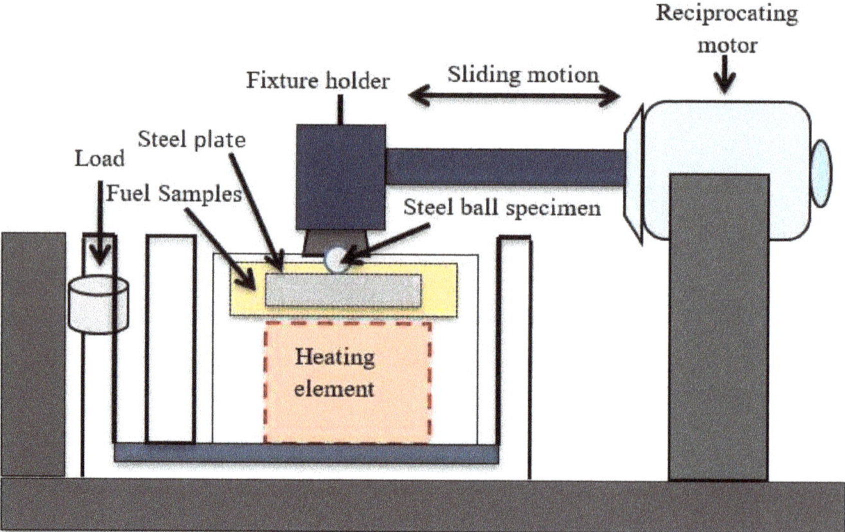

Figure 3. Schematic view of HFRR test rig.

2.3.3. Four-Ball Tribo Tester Rig

An automatic four-ball tribo tester (FBT-3, Ducom Instruments, Bengaluru, India) as shown in Figure 4 was used to study the effect of different fuels on lubricant tribological characteristics. Then, 10 mL of lubricant sample was poured into the cup holder containing three stationary steel balls attached to the temperature sensor. For each experiment, four new separate steel balls were used. Commercial lubricant SAE-40 was also tested as a reference lubricant for comparing the results. All experiments were performed according to the ASTM D4172 standard; the working conditions were as follows: test duration (60 min), applied load (40 kg), oil temperature (60 °C), and rotational speed of spindle (1200 rpm).

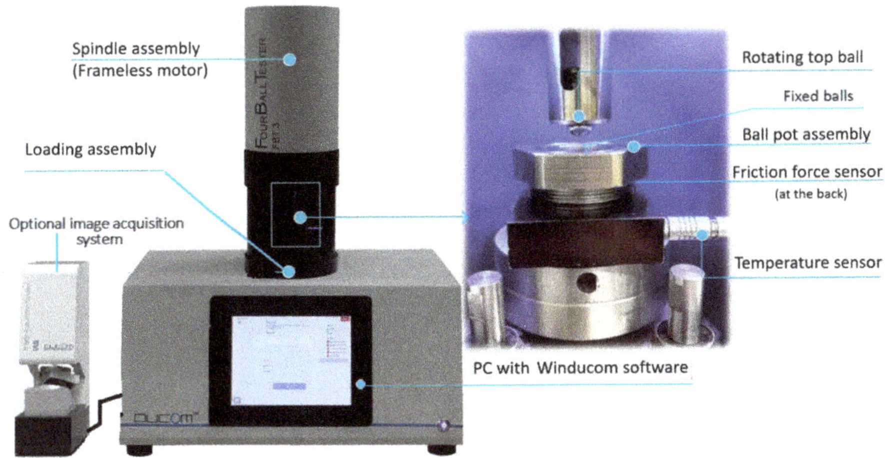

Figure 4. Schematic view of four-ball tribo tester rig.

3. Results and Discussion

3.1. Engine Performance

Brake-specific fuel consumption (BSFC) and BTE results for fuel samples at full load condition with variable engine speed are presented in Figure 5a,b. The BSFC values of various tested fuel samples, i.e., B10, B30, B30 + TiO$_2$, and B30 + DMC, were 0.373 kg/kWh, 0.384 kg/kWh, 0.403 kg/kWh, and 0.407 kg/kWh, respectively, at 2050 rpm are shown in Figure 5a. On average, the B30 + TiO$_2$ ternary blend exhibited the lowest BSFC among all the tested fuels. The average BSFC value of B30 + DMC decreased compared to B30 but increased in comparison to B10 due to a lower heating value [28]. Ternary fuel blends showed improvements in BSFC reduction of 6.76% and 1.45% for B30 + TiO$_2$ and B30 + DMC in comparison with B30. The lower BSFC in the case of an alcoholic ternary blend was due to lower viscosity and density which improved the fuel spray characteristics and led to a better A:F mixture due to enhanced atomization [29,30]. The higher evaporation rate of alcoholic ternary blends resulted in enhanced combustion properties and better fuel spray characteristics. The nanoparticle ternary fuel blend showed a significant reduction in BSFC due to a lower ignition delay that led to less premixed combustion of air and fuel mixture [31]. The nanoparticles as fuel improvers avoided the deposition of carbon and iron particles, resulting in decreased friction among diesel engine components, thus leading to an increase in BP and torque along with a reduction in BSFC.

Figure 5b exhibits the effect of speed on BTE. On average, 8.01% and 5.49% increments in BTE values were noted for the B30 blend with additives DMC and TiO$_2$, respectively, in comparison with neat B30 fuel. There was an enhancement in the BTE value when the alcohol fuel additive was added, which acted as an oxygen donor due to a reduction in combustion time and enhanced combustion process [32]. The thermal efficiency of diesel–biodiesel–fuel additive blends improved due to improved combustion with a supply of excess oxygen in the fuel-rich zone in the compression ignition engine combustion chamber [30,33]. TiO$_2$ nanoparticles enhanced the density of fuel–air charge due to the high evaporation rate of fuel which resulted in higher power and BTE [34]. A higher BTE value was observed because of the addition of titanium oxide nanoparticles as a nano fuel additive, which have a high chemical reactive surface and surface volume ratio, thereby boosting the combustion by proving better oxidation. Similar results reported by various researchers. Örs et al. [35] reported a 24.52% improvement in the BSFC of B20 blend with the addition of TiO$_2$ compared to B20 and the average reduction in the BSFC for B20 with TiO$_2$ was 27.73%. Silva et al. [36] also found a 21.28% reduction in BSFC due to the addition of TiO$_2$ nanoparticles compared to petroleum diesel. Yuvarajan Devarajan [37] showed a 4.1% reduction in BSFC and a 1.6% increment in BTE with the addition of 20% DMC to biodiesel blends.

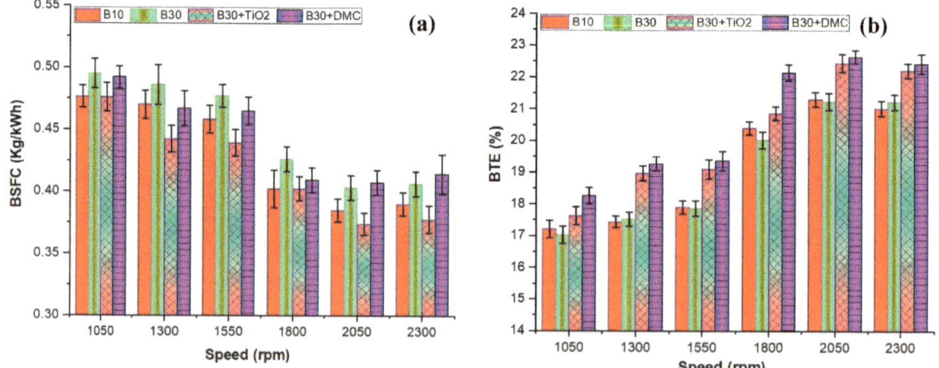

Figure 5. Variation of engine performance parameters: (**a**) brake-specific fuel consumption (BSFC) and (**b**) brake thermal efficiency (BTE) with respect to speed at 100% engine load.

3.2. Engine Emissions

Exhaust emission (HC, NOx, and CO) results for tested fuel samples are exhibited in Figure 6a–c. The formation of CO emissions is mainly dependent on the combustion process. Incomplete combustion leads to CO formation. The presence of oxygen molecules in fuel assists in the completion of the combustion process and conversion of CO to CO_2. From Figure 6a, it is evident that CO emissions reduced significantly in the case of biodiesel–diesel and for the alcohol and nano fuel blends in comparison to the standard commercially used Malaysian (B10) diesel. All ternary blends showed an average significant reduction in CO emissions with the addition of fuel additives, with 32.09% and 12.46% for B30 with the fuel additives DMC and TiO_2, respectively, in comparison with the B30 blend. In another study conducted by Örs et al. [35], it was observed that the addition of TiO_2 reduced the CO emissions by 10.83% and 25.56% for B20 with TiO_2 compared to B20 and petroleum diesel, respectively. The same observation was found by Saxena et al. [38]. A 7.4% reduction in CO emissions was reported by Yuvarajan Devarajan [37] with the addition of 20% DMC to the fuel blend. The oxygenated alcohol ternary blend showed a remarkable reduction in CO emissions due to the presence of a high amount of oxygen content compared to other fuel samples. Nanoparticle ternary fuel blends also reduced CO emissions due to the potential redox active property that assisted in the complete conversion of CO to CO_2 [39]. The chemical reactive surface of nanoparticles due to a higher surface volume reduces the ignition delay, resulting in a better combustion process and consequently reducing CO emissions [40].

Figure 6b exhibits the trend of HC emissions with respect to engine speed. Similarly, incomplete combustion is the source of HC emissions. The reduction in HC emissions was due to high oxygen content and a higher cetane number of ternary fuel blends, which resulted in lower ignition delay [41]. All ternary blends and the B30 fuel sample showed a substantial decline in hydrocarbon emissions in comparison to B10. B30 fuel blended with DMC and titanium dioxide fuel additives showed reductions in HC emissions by 25.4% and 8.63%, respectively, in comparison with the neat B30 fuel, due to higher oxygen content. An average reduction in HC emissions of 34.12% for B20 with TiO_2 was reported by Örs et al. [35]. In another study [37], the addition of DMC (20%) as a fuel additive minimized the HC emissions by up to 5.2%. The larger surface area of nanoparticles and higher oxygen amount in ternary fuel blends improved the fuel combustion process, which led to lower HC emissions. Oxygenated ternary fuel blends increased the in-cylinder pressure and temperature and heat release rate, which resulted in the improved combustion process and lower HC emissions [42]. Nanoparticles accelerated the oxidation process of hydrocarbons into CO_2 and water, consequently reducing HC emissions [43].

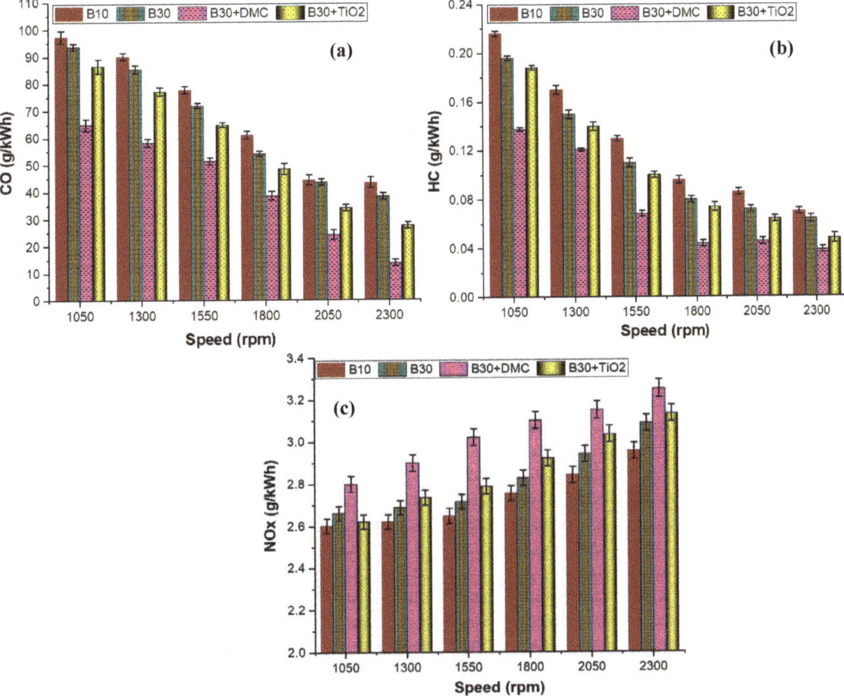

Figure 6. Variation of engine exhaust emissions: (**a**) CO, (**b**) HC, and (**c**) NOx with respect to speed at 100% engine load.

Figure 6c represents the NOx emissions for all tested fuel samples. Ternary fuel samples showed an increment in NOx emissions compared to B10 and B30. The test fuel blends demonstrated an increase in NOx emissions by 9.72% and 1.84% for B30 fuel blended with DMC and titanium dioxide fuel additives in comparison with the neat B30 fuel. The higher oxygen content and cetane number of ternary fuel blends increased the in-cylinder temperature and pressure, which led to higher NOx emissions. A similar increment in NOx emissions was reported in B20 with TiO_2 up to 6.95% compared to B20 [35]. Significant increments in NOx emissions were observed with the addition of DMC (10% and 20%) to fuel blends compared to petroleum diesel [42]. The oxygenated ternary fuel blend contained a higher amount of oxygen content among all tested fuels, which resulted in high in-cylinder combustion temperature and pressure, consequently increasing NOx emissions [44]. Nanoparticle ternary fuel blends improved the combustion process by providing excess oxygen, which resulted in a high combustion temperature and higher NOx. The higher thermal conductivity and large surface area of nanoparticles increased the in-cylinder pressure and enhanced the combustion process, resulting in higher NOx emissions [40].

3.3. HFRR Tribological Study

The COF and WSD results of all five tested fuel samples are exhibited in Figure 7a,b. The lubricity of fuel is a very important factor that should be considered before the application of fuel in the automotive industry. Engine life is mainly dependent on the lubricity of the fuel. Diesel engine components (fuel pump, fuel injector, etc.) are self-lubricated with the fuel itself. The COF trend with respect to time is presented in Figure 7a. During the early stage of the experiment, all tested fuels showed a sharp rise in COF for a few minutes, named the run-in period. During this period, there is no lubricating film between mating surfaces, leading to very high COF. After this run-in period, steady-state conditions are achieved due to the formation of a thin lubricating protective layer between metallic surfaces. B10 (Malaysian commercial diesel) showed a higher COF value than all fuel samples due to a lower percentage of ester molecules compared to other fuel samples. The unsteady state for B10 (Malaysian commercial diesel) increased with time compared to other samples because B30 with fuel additives had a shorter run-in period due to the presence of ester molecules and nanoparticles that created a protective layer quickly compared to B10. The ester molecules in fuel samples assisted in the formation of the protective lubricating film between mating surfaces [45]. The pure biodiesel fuel sample showed a minimum run-in period due to the adsorption of ester molecules on the metallic surface, which acted as a protective layer during the rubbing process. B30 and B30 with fuel additives showed a significant reduction in COF values compared to B10 because of ester molecules and nanoparticles which acted as a protective layer between mating surfaces. From Figure 7b, it is evident that pure biodiesel showed the lowest average COF value among all tested fuel samples. The higher percentage of unsaturated fatty acids in the biodiesel sample improved the lubricity of fuel and resulted in lower COF and WSD. Among all tested fuel blends, B30 + TiO_2 showed a significant reduction in COF due to the presence of spherical-shape nanoparticles that acted as a surfactant between metallic surfaces. Nanoparticles also acted as a ball bearing between rubbing surfaces, consequently reducing COF and WSD. The alcoholic ternary blend exhibited the highest COF value amongst all the fuel blends except for B10. B30 blended with DMC fuel contained a higher amount of oxygen content, which led to the lowest WSD among all tested fuels due to the formation of an anti-adhesive oxide layer between metallic surfaces at the initial stage of the experiment. On average, B30 + TiO_2 showed a reduction of 6.72% and B30 + DMC exhibited an increment of 7.15% in COF in comparison to B30 fuel. Both ternary blends showed a significant reduction in WSD by 38.4% and 23.5% for B30 fuel blended with DMC and TiO_2 fuel additives in comparison with the neat B30 fuel blend.

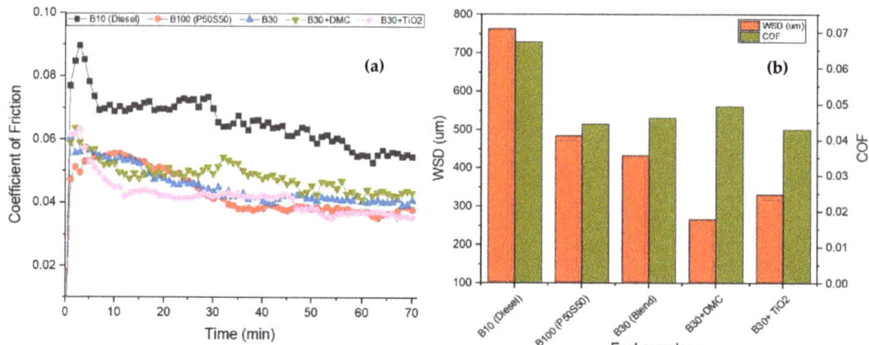

Figure 7. Coefficient of friction (**a**)(COF) and (**b**)wear scar diameter (WSD) trend for all tested fuel samples.

3.4. Four-Ball Tribological Study

The effect of different fuel samples on the lubricity of mineral lubricant is presented in Figure 8a,b. According to previous literature, the lubricant is contaminated with fuel up to 5% due to crankcase dilution [20]. The lubricant was contaminated with combustible fuel which altered the tribological properties of lubricant and resulted in poor lubricating characteristics due to lubricant degradation. Figure 8a exhibits the friction coefficient trend for all tested samples. The run-in period of pure mineral lubricant was much lower due to better lubricating characteristics. Steady-state conditions were achieved quickly due to the formation of the lubricating film between metallic contacts in the initial phase of the experimental run [46]. The mineral lubricant showed a much lower COF compared to other contaminated samples of lubricant with different fuels. The addition of combustible fuel to the lubricant altered its lubricating properties and resulted in poor tribological behavior. It is also evident from Figure 8a that mineral lubricant contaminated with a higher percentage of biodiesel fuel sample (B30) and B30 with fuel additives showed less lubricant degradation compared to B10. The lubricant + B10 sample showed the highest COF value among all tested samples because of a higher percentage of diesel in the sample compared to other tested samples. All contaminated lubricant samples showed high COF due to a decrease in viscosity, which had a high influence on the lubricity of fuel compared to mineral lubricant. Among ternary contaminated samples, Lub + B30 + TiO$_2$ showed the best friction coefficient due to the presence of nanoparticles which acted as a surface between metallic contacts and resulted in lower COF compared to other contaminated lubricant samples. Figure 8b presents the average results of COF and WSD for all tested samples. Lub + B10 showed the highest COF value among all tested samples due to the presence of high sulfur content and a low percentage of ester molecules, which resulted in a poor lubricating film between metallic contacts, consequently increasing COF and WSD. On average, Lub + B30 + TiO$_2$ exhibited the lowest COF value among all contaminated lubricant samples due to the presence of spherical-shape nanoparticles which acted as a friction-reducing agent between rubbing surfaces. Lub + B30 also showed a lower COF compared to Lub + B10 due to the presence of a high amount of ester molecules. Yashvir Singh et al. [25] also found that more than 5–8% contamination of 100% pure moringa biodiesel fuel in lubricant enhanced the lubricity, but more than 8% contamination of moringa biodiesel fuel increased the wear rate considerably. Similarly, Maleque et al. [23] and Sulek et al. [24] reported that 5% biodiesel fuel dilution in commercial lubricant decreased the wear rate. All contaminated samples showed increments in COF of 13.72%, 27%, 31.35%, and 42.29% for Lub + B30 + TiO$_2$, Lub + B30 + DMC, and Lub + B10, respectively, compared to mineral lubricant. All contaminated lubricant samples showed lower WSD values compared to mineral lubricant due to the presence of ester molecules, long-chain carbon atoms, and fuel additives (oxygenated alcohols and nanoparticles), which acted as a surfactant between metallic contacts. On average, WSD values decreased by 28.9%, 25.24%, 24.25%, and 17% for Lub + B30, Lub + B30 + DMC, Lub + B30 + TiO$_2$, and Lub + B10, respectively, compared to mineral lubricant.

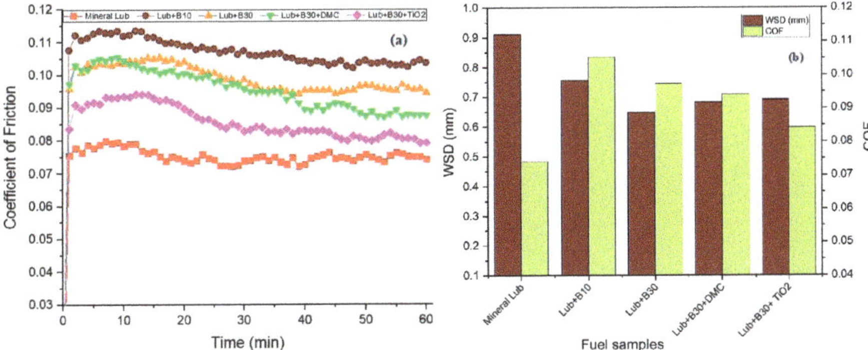

Figure 8. Tribological behavior: (**a**) COF and (**b**) WSD of mineral lubricant and lubricant contaminated samples.

4. Conclusions

In this current study, palm–sesame biodiesel was produced, and various biodiesel–diesel–fuel additive blends were prepared to examine their lubricity, as well as the effect of these blends on the contamination of lubricant and the effect of ternary fuel blends on diesel engine characteristics. Based on the results obtained, the conclusions below were made.

1. B10 (Malaysian commercial diesel) presented very poor lubricity. The B10 fuel blend showed very high COF and WSD values compared to other tested fuels.
2. On average, B30 + TiO$_2$ showed a reduction of 6.72% and B30 + DMC exhibited an increment of 7.15% in COF in comparison to B30 fuel. Both fuel additive blends showed a significant reduction in WSD by 38.4% and 23.5% for B30 fuel blended with DMC and TiO$_2$ fuel additives in comparison with the neat B30 fuel blend.
3. All contaminated samples showed increments in COF if 13.72%, 27%, 31.35%, and 42.29% for Lub + B30 + TiO$_2$, Lub + B30 + DMC, and Lub + B10, respectively, compared to mineral lubricant.
4. Ternary test fuels demonstrated an improvement in BSFC reduction of 6.76% and 1.45% for B30 fuel blended with DMC and TiO$_2$ fuel additives in comparison with the neat B30 fuel blend.
5. All ternary test fuels demonstrated a significant reduction in emissions of carbon monoxide upon adding fuel additives by 32.09% and 12.46% for B30 fuel blended with DMC and TiO$_2$, respectively, in comparison with the B30 fuel blend.
6. The B30 fuel blended with DMC and TiO$_2$ nanoparticles showed a reduction in HC emissions of 25.4% and 8.63%, respectively, compared to B30 due to the presence of higher oxygen content.
7. On average, the blends with fuel additives resulted in an increase in NOx emissions by 9.72% and 1.84% for B30 fuel blended with DMC and TiO$_2$ in comparison with the B30 fuel blend.

The main findings of this current research work are that Malaysian commercial diesel showed poor tribological characteristics, while the nanoparticle (TiO$_2$) blended fuel showed the best lubricity among all tested fuel samples. The oxygenated alcohol (DMC) blended fuel showed better engine performance and emission (HC and CO) characteristics among all tested fuel samples.

5. Future Recommendations

Finally, we suggest that nanoparticles as fuel additives are more feasible due to their promising results. TiO$_2$ nanoparticles showed a significant reduction in COF compared to oxygenated alcoholic fuel additives. TiO$_2$ nanoparticles acted as a friction-reducing agent due to their spherical shape, which established a thin lubricating film between metallic contacts, consequently reducing COF. In the future, a long endurance test should be conducted to investigate the effect of fuel additives and lubricant contamination on the diesel engine components.

Author Contributions: Conceptualization, M.A.M and M.A.K.; methodology, M.A.M.; validation, L.R., F.N., and M.F.; formal analysis, M.G.; investigation, M.A.M., S.B., and M.E.M.S.; writing—original draft preparation, M.A.M.; writing—review and editing, I.M.R.F. and H.C.O.; supervision, H.H.M. All authors read and agreed to the published version of the manuscript.

Funding: The authors take the opportunity to thank the UNIVERSITY OF MALAYA for the financial support under research grant FP142-2019A under the FRGS from the University of Malaya, Kuala Lumpur, Malaysia, and HEC ISLAMABAD, PAKISTAN for supporting the research under Grant No. 5-1/HRD/UESTPI (Batch-VI)/4954/2018/HEC.

Acknowledgments: The authors also acknowledge the contribution of the research development fund of the School of Information, Systems, and Modeling, University of Technology Sydney, Australia.

Conflicts of Interest: The authors declare no conflict of interest.

Nomenclature

ASTM	American Standard for Testing Materials
B100	100% biodiesel
EN	Europe Union
CI	Compression ignition
B30	70% diesel + 30% biodiesel

KOH	Potassium hydroxide
B10	10% biodiesel + 90% diesel
BTE	Brake thermal efficiency
HFRR	High-frequency reciprocating rig
BSFC	Brake-specific fuel consumption
TiO_2	Titanium oxide
BTDC	Before top dead center
P50S50	50% palm and 50% sesame
SO	Sesame oil
B30 + TiO_2	70% diesel + 30% biodiesel + 100 ppm TiO_2 by mass
COF	Coefficient of friction
ppm	Parts per million
HC	Hydrocarbon
DMC	Dimethyl carbonate
CO	Carbon monoxide
WSD	Wear scar diameter
NO_x	Nitrogen oxides

References

1. Fattah, I.M.R.; Ong, H.C.; Mahlia, T.M.I.; Mofijur, M.; Silitonga, A.S.; Rahman, S.M.A.; Ahmad, A. State of the Art of Catalysts for Biodiesel Production. *Front. Energy Res.* **2020**, *8*. [CrossRef]
2. Ong, H.C.; Milano, J.; Silitonga, A.S.; Hassan, M.H.; Shamsuddin, A.H.; Wang, C.-T.; Indra Mahlia, T.M.; Siswantoro, J.; Kusumo, F.; Sutrisno, J. Biodiesel production from Calophyllum inophyllum-Ceiba pentandra oil mixture: Optimization and characterization. *J. Clean. Prod.* **2019**, *219*, 183–198. [CrossRef]
3. Silitonga, A.; Shamsuddin, A.; Mahlia, T.; Milano, J.; Kusumo, F.; Siswantoro, J.; Dharma, S.; Sebayang, A.; Masjuki, H.; Ong, H.C. Biodiesel synthesis from Ceiba pentandra oil by microwave irradiation-assisted transesterification: ELM modeling and optimization. *Renew. Energy* **2020**, *146*, 1278–1291. [CrossRef]
4. Ong, H.C.; Masjuki, H.H.; Mahlia, T.M.I.; Silitonga, A.S.; Chong, W.T.; Yusaf, T. Engine performance and emissions using Jatropha curcas, Ceiba pentandra and Calophyllum inophyllum biodiesel in a CI diesel engine. *Energy* **2014**, *69*, 427–445. [CrossRef]
5. Gad, M.S.; Jayaraj, S. A comparative study on the effect of nano-additives on the performance and emissions of a diesel engine run on Jatropha biodiesel. *Fuel* **2020**, *267*, 117168. [CrossRef]
6. Mahlia, T.M.I.; Syazmi, Z.A.H.S.; Mofijur, M.; Abas, A.E.P.; Bilad, M.R.; Ong, H.C.; Silitonga, A.S. Patent landscape review on biodiesel production: Technology updates. *Renew. Sustain. Energy Rev.* **2020**, *118*, 109526. [CrossRef]
7. Silitonga, A.S.; Masjuki, H.H.; Mahlia, T.M.I.; Ong, H.C.; Chong, W.T.; Boosroh, M.H. Overview properties of biodiesel diesel blends from edible and non-edible feedstock. *Renew. Sustain. Energy Rev.* **2013**, *22*, 346–360. [CrossRef]
8. Imtenan, S.; Masjuki, H.H.; Varman, M.; Rizwanul Fattah, I.M.; Sajjad, H.; Arbab, M.I. Effect of n-butanol and diethyl ether as oxygenated additives on combustion–emission-performance characteristics of a multiple cylinder diesel engine fuelled with diesel–jatropha biodiesel blend. *Energy Convers. Manag.* **2015**, *94*, 84–94. [CrossRef]
9. Fattah, I.M.R.; Masjuki, H.H.; Kalam, M.A.; Wakil, M.A.; Ashraful, A.M.; Shahir, S.A. Experimental investigation of performance and regulated emissions of a diesel engine with Calophyllum inophyllum biodiesel blends accompanied by oxidation inhibitors. *Energy Convers. Manag.* **2014**, *83*, 232–240. [CrossRef]
10. Latiff, R. *Malaysia to Implement B30 Biodiesel Mandate in Transport Sector before 2025*; Sarkar, H., Ed.; REUTERS: Putrajaya, Malaysia, 2020.
11. Elisha, O.; Fauzi, A.; Anggraini, E. Analysis of Production and Consumption of Palm-Oil Based Biofuel using System Dynamics Model: Case of Indonesia. *Int. J. Sci. Eng. Technol.* **2019**, *6*. [CrossRef]
12. Coca, N. *As Palm Oil for Biofuel Rises in Southeast Asia, Tropical Ecosystems Shrink*; Chinadialogue China: Beijing, China, 2020.

13. Fattah, I.M.R.; Masjuki, H.H.; Kalam, M.A.; Mofijur, M.; Abedin, M.J. Effect of antioxidant on the performance and emission characteristics of a diesel engine fueled with palm biodiesel blends. *Energy Convers. Manag.* **2014**, *79*, 265–272. [CrossRef]

14. Mujtaba, M.A.; Muk Cho, H.; Masjuki, H.H.; Kalam, M.A.; Ong, H.C.; Gul, M.; Harith, M.H.; Yusoff, M.N.A.M. Critical review on sesame seed oil and its methyl ester on cold flow and oxidation stability. *Energy Rep.* **2020**, *6*, 40–54. [CrossRef]

15. Mujtaba, M.A.; Masjuki, H.H.; Kalam, M.A.; Ong, H.C.; Gul, M.; Farooq, M.; Soudagar, M.E.M.; Ahmed, W.; Harith, M.H.; Yusoff, M.N.A.M. Ultrasound-assisted process optimization and tribological characteristics of biodiesel from palm-sesame oil via response surface methodology and extreme learning machine-Cuckoo search. *Renew. Energy* **2020**. [CrossRef]

16. Gul, M.; Masjuki, H.H.; Kalam, M.A.; Zulkifli, N.W.M.; Mujtaba, M.A. A Review: Role of Fatty Acids Composition in Characterizing Potential Feedstock for Sustainable Green Lubricants by Advance Transesterification Process and its Global as Well as Pakistani Prospective. *BioEnergy Res.* **2020**, *13*, 1–22. [CrossRef]

17. Lapuerta, M.; García-Contreras, R.; Agudelo, J.R. Lubricity of Ethanol-Biodiesel-Diesel Fuel Blends. *Energy Fuels* **2010**, *24*, 1374–1379. [CrossRef]

18. Liaquat, A.M.; Masjuki, H.H.; Kalam, M.A.; Rizwanul Fattah, I.M. Impact of biodiesel blend on injector deposit formation. *Energy* **2014**, *72*, 813–823. [CrossRef]

19. Dhar, A.; Agarwal, A.K. Experimental investigations of effect of Karanja biodiesel on tribological properties of lubricating oil in a compression ignition engine. *Fuel* **2014**, *130*, 112–119. [CrossRef]

20. Arumugam, S.; Sriram, G. Preliminary study of nano-and microscale TiO2 additives on tribological behavior of chemically modified rapeseed oil. *Tribol. Trans.* **2013**, *56*, 797–805. [CrossRef]

21. Agarwal, A.K. Lubricating Oil Tribology of a Biodiesel-Fuelled Compression Ignition Engine. Proceedings of ASME 2003 Internal Combustion Engine Division Spring Technical Conference, Erie, PA, USA, 7–10 September 2003.

22. Gul, M.; Zulkifli, N.W.M.; Masjuki, H.H.; Kalam, M.A.; Mujtaba, M.A.; Harith, M.H.; Syahir, A.Z.; Ahmed, W.; Bari Farooq, A. Effect of TMP-based-cottonseed oil-biolubricant blends on tribological behavior of cylinder liner-piston ring combinations. *Fuel* **2020**, *278*, 118242. [CrossRef]

23. Maleque, M.A.; Masjuki, H.H.; Haseeb, A.S.M.A. Effect of mechanical factors on tribological properties of palm oil methyl ester blended lubricant. *Wear* **2000**, *239*, 117–125. [CrossRef]

24. Sulek, M.; Kulczycki, A.; Malysa, A. Assessment of lubricity of compositions of fuel oil with biocomponents derived from rape-seed. *Wear* **2010**, *268*, 104–108. [CrossRef]

25. Singh, Y.; Singla, A.; Upadhyay, A.; Singh, A.K. Sustainability of Moringa-oil–based biodiesel blended lubricant. *Energy Sources Part A: Recovery Util. Environ. Eff.* **2017**, *39*, 313–319. [CrossRef]

26. Arumugam, S.; Sriram, G. Effect of Bio-Lubricant and Biodiesel-Contaminated Lubricant on Tribological Behavior of Cylinder Liner–Piston Ring Combination. *Tribol. Trans.* **2012**, *55*, 438–445. [CrossRef]

27. Mofijur, M.; Kusumo, F.; Fattah, I.; Mahmudul, H.; Rasul, M.; Shamsuddin, A.; Mahlia, T. Resource Recovery from Waste Coffee Grounds Using Ultrasonic-Assisted Technology for Bioenergy Production. *Energies* **2020**, *13*, 1770. [CrossRef]

28. Mofijur, M.; Masjuki, H.H.; Kalam, M.A.; Atabani, A.E.; Fattah, I.M.R.; Mobarak, H.M. Comparative evaluation of performance and emission characteristics of Moringa oleifera and Palm oil based biodiesel in a diesel engine. *Ind. Crop. Prod.* **2014**, *53*, 78–84. [CrossRef]

29. Murcak, A.; Haşimoğlu, C.; Çevik, İ.; Karabektaş, M.; Ergen, G. Effects of ethanol–diesel blends to performance of a DI diesel engine for different injection timings. *Fuel* **2013**, *109*, 582–587. [CrossRef]

30. Imtenan, S.; Masjuki, H.H.; Varman, M.; Rizwanul Fattah, I.M. Evaluation of n-butanol as an oxygenated additive to improve combustion-emission-performance characteristics of a diesel engine fuelled with a diesel-calophyllum inophyllum biodiesel blend. *RSC Adv.* **2015**, *5*, 17160–17170. [CrossRef]

31. Nadeem, M.; Rangkuti, C.; Anuar, K.; Haq, M.; Tan, I.; Shah, S. Diesel engine performance and emission evaluation using emulsified fuels stabilized by conventional and gemini surfactants. *Fuel* **2006**, *85*, 2111–2119. [CrossRef]

32. Khalife, E.; Tabatabaei, M.; Demirbas, A.; Aghbashlo, M. Impacts of additives on performance and emission characteristics of diesel engines during steady state operation. *Prog. Energy Combust. Sci.* **2017**, *59*, 32–78. [CrossRef]

33. Imtenan, S.; Varman, M.; Masjuki, H.H.; Kalam, M.A.; Sajjad, H.; Arbab, M.I.; Fattah, I.M.R. Impact of low temperature combustion attaining strategies on diesel engine emissions for diesel and biodiesels: A review. *Energy Convers. Manag.* **2014**, *80*, 329–356. [CrossRef]

34. Anbarasu, A.; Karthikeyan, A. Performance and emission characteristics of a diesel engine using cerium oxide nanoparticle blended biodiesel emulsion fuel. *J. Energy Eng* **2016**, *142*, 04015009. [CrossRef]

35. Örs, I.; Sarıkoç, S.; Atabani, A.E.; Ünalan, S.; Akansu, S.O. The effects on performance, combustion and emission characteristics of DICI engine fuelled with TiO2 nanoparticles addition in diesel/biodiesel/n-butanol blends. *Fuel* **2018**, *234*, 177–188. [CrossRef]

36. D'Silva, R.; Binu, K.; Bhat, T.J.M.T.P. Performance and Emission characteristics of a CI Engine fuelled with diesel and TiO$_2$ nanoparticles as fuel additive. *Mater. Today Proc.* **2015**, *2*, 3728–3737. [CrossRef]

37. Devarajan, Y. Experimental evaluation of combustion, emission and performance of research diesel engine fuelled di-methyl- carbonate and biodiesel blends. *Atmos. Pollut. Res.* **2019**, *10*, 795–801. [CrossRef]

38. Saxena, V.; Kumar, N.; Saxena, V.K. Multi-objective optimization of modified nanofluid fuel blends at different TiO2 nanoparticle concentration in diesel engine: Experimental assessment and modeling. *Appl. Energy* **2019**, *248*, 330–353. [CrossRef]

39. Prabu, A.; Anand, R.B. Emission control strategy by adding alumina and cerium oxide nano particle in biodiesel. *J. Energy Inst.* **2016**, *89*, 366–372. [CrossRef]

40. Hoseini, S.S.; Najafi, G.; Ghobadian, B.; Ebadi, M.T.; Mamat, R.; Yusaf, T. Biodiesels from three feedstock: The effect of graphene oxide (GO) nanoparticles diesel engine parameters fuelled with biodiesel. *Renew. Energy* **2020**, *145*, 190–201. [CrossRef]

41. Anand, R.; Kannan, G.; Nagarajan, S.; Velmathi, S. Performance emission and combustion characteristics of a diesel engine fueled with biodiesel produced from waste cooking oil. *SAE Tech. Pap.* **2010**. [CrossRef]

42. Pan, M.; Qian, W.; Zheng, Z.; Huang, R.; Zhou, X.; Huang, H.; Li, M. The potential of dimethyl carbonate (DMC) as an alternative fuel for compression ignition engines with different EGR rates. *Fuel* **2019**, *257*, 115920. [CrossRef]

43. Kalam, M.A.; Masjuki, H.H. Testing palm biodiesel and NPAA additives to control NOx and CO while improving efficiency in diesel engines. *Bioenergy* **2008**, *32*, 1116–1122. [CrossRef]

44. Atmanli, A. Comparative analyses of diesel–waste oil biodiesel and propanol, n-butanol or 1-pentanol blends in a diesel engine. *Fuel* **2016**, *176*, 209–215. [CrossRef]

45. Konishi, T.; Klaus, E.E.; Duda, J.L. Wear Characteristics of Aluminum-Silicon Alloy under Lubricated Sliding Conditions. *Tribol. Trans.* **1996**, *39*, 811–818. [CrossRef]

46. Habibullah, M.; Masjuki, H.H.; Kalam, M.A.; Zulkifli, N.W.M.; Masum, B.M.; Arslan, A.; Gulzar, M. Friction and wear characteristics of Calophyllum inophyllum biodiesel. *Ind. Crop. Prod.* **2015**, *76*, 188–197. [CrossRef]

MDPI

St. Alban-Anlage 66

4052 Basel

Switzerland

Tel. +41 61 683 77 34

Fax +41 61 302 89 18

www.mdpi.com

Energies Editorial Office

E-mail: energies@mdpi.com

www.mdpi.com/journal/energies